COMPOSITE BASICS

SEVEN

Seventh Edition

Andrew C. Marshall

Seventh Edition, first printing, June 2005

Layout, Design & Typography
by Marilyn McCracken
Phone 562-201-1429
Whittier, CA

Published by Marshall Consulting
720 Appaloosa Drive, Walnut Creek, CA 94596
Phone: 925-945-6051 • Fax: 925-945-1461

Printed in the United States of America

ISBN: 0-9664540-4-9

ABOUT THE AUTHOR

Andrew C. Marshall has an international reputation for his role in the development and application of core materials, sandwich structures and composites in both the aerospace and industrial fields.

He is currently the president of his own consulting firm, whose clients have included companies such as Boeing Aerospace, Boeing Vertol, British Aerospace, Dupont, Gibbs and Cox, Glastron Boats, Hexcel, Israeli Aircraft, Nippon Graphite Fiber, Nippon Oil, Rutan Aircraft, Voyager Aircraft, Bay Area Rapid Transit (BART) and many others.

Mr. Marshall has authored many technical papers, articles and manuals in this field. These include a chapter in *Handbook of Composites*, published in 1982 by Van Nostrand Reinhold, a Second Edition published in 1997 by Chapman & Hall, a section on *Core Composites* in the *International Encyclopedia of Composites* published in 1990 by VCH Publishers.

In addition he wrote a series of 28 magazine articles, titled *Composite Basics,* which appeared in the magazine, *Homebuilt Aircraft*, and were then rewritten to form the First Edition of this book.

Mr. Marshall first soloed in 1936, obtained his Private Pilot's License and Limited Commercial License in 1941, and has since logged more than 1800 total hours in many small airplanes.

A graduate of the University of California (Berkeley) in Mechanical Engineering, he is also a Fellow of the Society for the Advancement of Materials and Process Engineering, a former Vice President of the Hexcel Corporation and a Registered Professional Engineer in the State of California.

PREFACE to the SEVENTH EDITION

The original manuscript of this book was compiled from a series of articles which appeared in the magazine, *HOMEBUILT AIRCRAFT,* between 1983 and 1985. The first edition was then edited and published in 1985, the second in 1989, the third in 1993, the fourth in 1994, the fifth in 1998, and the sixth in 2001.

The acceptance of the book still continues to be somewhat startling, with press runs selling out in much less time than anticipated. This seventh edition was prepared in 2005. The organization of the subject material is nearly the same as in the previous editions, with even more attention paid to assisting the reader in being able to find easily a specific item of interest.

The original style and character of the previous editions have been carefully preserved in this seventh edition, although, again, some new material has been added, and some of the text altered. It is hoped that this new version will be as quickly accepted and digested by new entries in the field of composites as has been the case with earlier editions.

Comments and suggestions from readers are solicited.

Andrew C. Marshall
925-945-6051

COMPOSITE BASICS - SEVEN

Seventh Edition

TABLE of CONTENTS

CHAPTER 1

THE STRUCTURAL FIBERS

Advanced composites are the most recent of the many new structural materials that continue to appear on the aerospace, racing automobile, sporting goods and boating scenes in the past fifty years. Surprisingly, they really are not all that new. Many of the composites which we call "advanced" are based on plain old glass drawn into fibers, a technology which began in the 1940's.

COMPOSITES

"Composite" is a general term which means an assembly of dissimilar materials used together to enable them to do a job that the individual materials cannot do by themselves. Under this definition, even reinforced concrete is a composite. "Advanced composites" refers to that group of materials usually associated with military or commercial aerospace structures and using the newer materials. They are carefully engineered and developed, and are uniquely suited to carrying substantial loads in structural members which have surprisingly light weight.

These materials usually consist of a resin matrix—typically an epoxy, polycyanate, polyester, or vinyl ester—and a fiber reinforcement. The fiber reinforcement can be glass, Kevlar, carbon fiber, Nextel, quartz, boron, or any of a number of other fibers all of which are very small in diameter and very strong, imparting a high degree of strength to the resulting mixture. This added strength is so large that the performance of a composite as a structure is usually another order of magnitude when compared to the strength of the resins by themselves. Even so, the strength and stiffness of the resin matrix *does* affect the finished composite structure, and stronger resins, such as epoxies, usually yield a higher strength structure than one which employs a lower strength resin, such as a general purpose polyester.

GLASS FIBERS

Glass fibers are normally furnished as yarn, or roving (a very large yarn). The properties of the composite structure are dependent on the amount, basic yarn strength, orientation and the construction of the yarn used. Usual construction of glass yarns is shown in Figure 1-1.

Glass fiber is produced in many different chemical compositions, each exhibiting somewhat different mechanical and chemical properties, and each designated by a different letter of the alphabet.

Of these many varieties, E-glass, C-glass and S-glass are the most commonly used. Figure 1-2 gives a comparison of the properties of these three materials. Of the three, E-glass is the most widely used, with S-glass used primarily in aircraft applications and C-glass appearing mostly in products designed to resist corrosion.

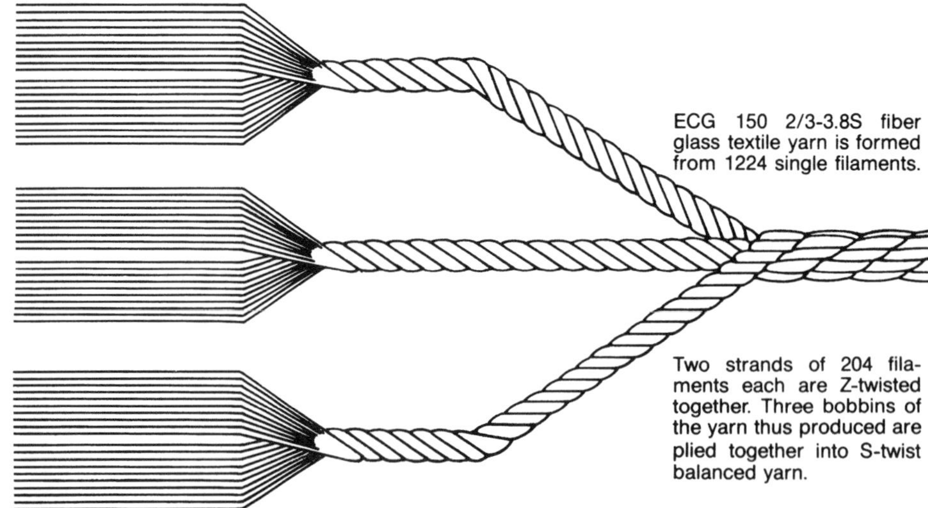

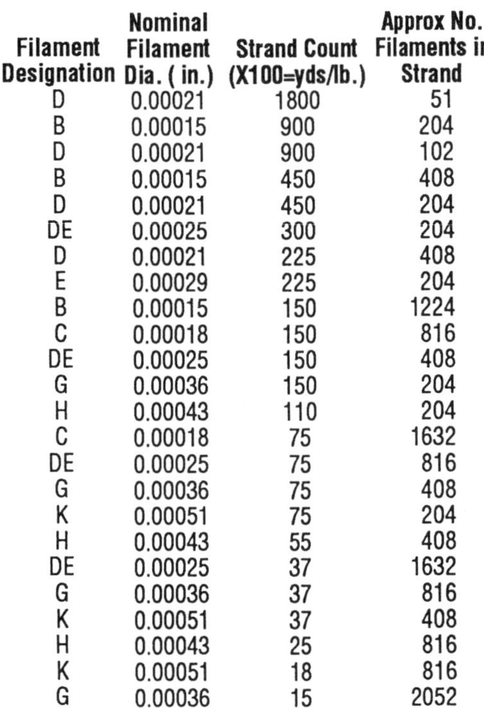

Filament Designation	Nominal Filament Dia. (in.)	Strand Count (X100=yds/lb.)	Approx No. Filaments in Strand
D	0.00021	1800	51
B	0.00015	900	204
D	0.00021	900	102
B	0.00015	450	408
D	0.00021	450	204
DE	0.00025	300	204
D	0.00021	225	408
E	0.00029	225	204
B	0.00015	150	1224
C	0.00018	150	816
DE	0.00025	150	408
G	0.00036	150	204
H	0.00043	110	204
C	0.00018	75	1632
DE	0.00025	75	816
G	0.00036	75	408
K	0.00051	75	204
H	0.00043	55	408
DE	0.00025	37	1632
G	0.00036	37	816
K	0.00051	37	408
H	0.00043	25	816
K	0.00051	18	816
G	0.00036	15	2052

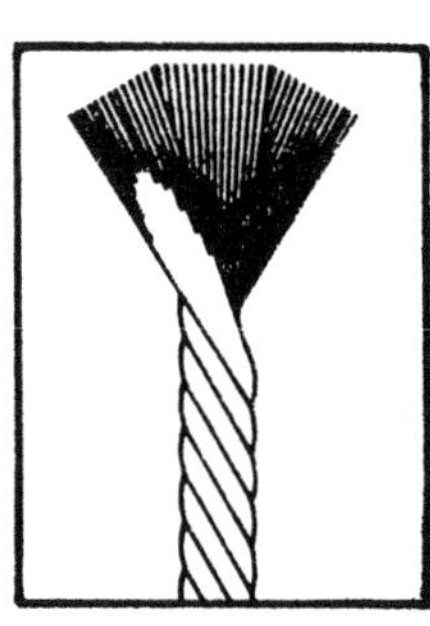

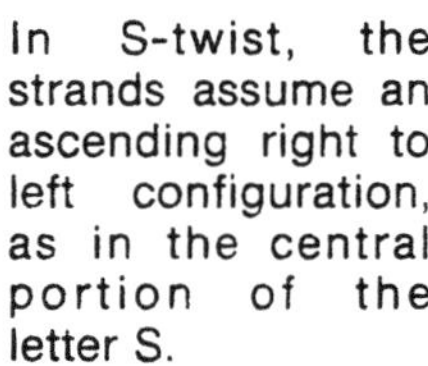

In S-twist, the strands assume an ascending right to left configuration, as in the central portion of the letter S.

In Z-twist, the strands assume an ascending left to right configuration, as in the central portion of the letter Z.

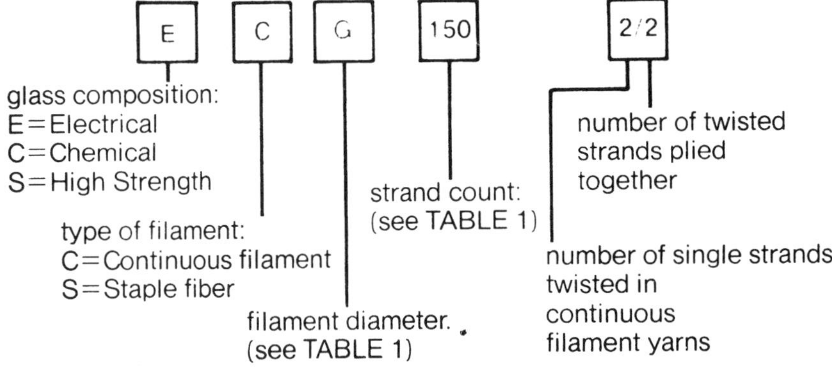

Figure 1.1. Glass Yarn Construction Details (Courtesy of PPG)

PROPERTY	GLASS TYPE E (Electrical)	C (Chemical)	S (High Strength)
PHYSICAL			
Specific Gravity	2.54	2.49	2.48
Moh Hardness	6.5	6.5	6.5
Contact Angle with Water, degress	0		
Coefficient of Friction with Glass	1.0		
Moisture Absorbency (surface), % up to	0.3		
Moisture Regain	none		
MECHANICAL			
Tensile strength, psi @ 72°F	500,000	480,000	665,000
@ 700°F	380,000		645.000
@1000°F	250,000		350,000
Tensile Modulus of Elasticity, psi @ 72°F	10.5 x 10^6	10.0 x 10^6	12.4 x 10^6
Hysteresis	none	none	none
Creep	none	none	none
Elongation at Break, %	4.8	4.8	5.7
Elastic Recovery, %	100	100	100
COMPOSITION			
Silicon oxide	54.3	64.6	64.2
Aluminum oxide	15.2	4.1	24.8
Ferrous oxide			0.21
Calcium oxide	17.2	13.2	0.01
Magnesium oxide	4.7	3.3	10.27
Sodium oxide	0.6	7.7	0.27
Potassium oxide		1.7	
Boron oxide	8.0	4.7	0.01
Barium oxide	0.9	0.9	0.2

Figure 1-2. Properties of Bare Single Filaments

In addition to these three glass formulations, several other special purpose glass fibers are available. These include M-glass, which is designed to have the highest possible elastic modulus, and L-glass, which has a high lead content and is sometimes used in the manufacture of composites intended to serve as radiation shielding.

Several new glass fiber types have also been developed to be competitive with the S-glass originally made by Owens Corning. These include R-glass and RH-Glass, both of which have been used by Gordon Plastics in the manufacture of "Bow-Tuf" and "Spar-Tuf", a precured unidirectional laminate available as wide as 12 inches in continuous lengths. Most glass fiber products, however, are still made from E glass. Unidirectional precured laminates are also offered by DFI Pultruded Composites, Inc.

In addition to the different formulations which make up the glass, the fibers themselves are quite different. For instance, a simple glass fiber which one may think of as being the primary ingredient in fiberglass fabric might be any one of about twenty-four different yarns. All of these are made from some seven different sized filaments ranging in diameter from 0.00013 inches (B fibers) to 0.00051 inches (K fibers). Figure 1-1 shows the diameters of those E-glass fibers commonly used in composites, as well as the system of the strand constructions used in yarns for weaving glass fabrics.

Each strand of yarn or roving contains a great many single filaments. Some of the heavier strands might carry as many as 2000 or more filaments in a single strand. Some of the lower cost materials such as Zoltek are offered in yarns containing as many as

40,000 individual filaments. These strands may then be twisted and plied so that the strand used to finally weave the cloth could be composed of two, three, six, or even more individual strands.

It is obvious that simply specifying that you want something made out of glass cloth does not really specify anything at all. The possible variations are mind-boggling.

Basically, glass fibers are made out of sand, gravel, clay, limestone, and other such materials which are melted down, with the quantities carefully adjusted to obtain the exact chemical formulation for the type of glass fiber desired. The molten glass is then drawn through tiny holes in platinum bushings to form the individual filaments of glass fiber. It is customary for the bushings to have a large number of holes in them so that when the fibers are drawn, a large number are produced all at the same time. All of the fibers in a single strand of glass are conventionally drawn as one group from the same pot of molten glass at the same time.

There are several companies who provide glass fibers on a commercial basis. The granddaddy of them all is the Owens Corning Fiberglass Co., but they now have spirited competition from firms such as PPG, Johns Manville, Certainteed and several others. The end result is that we have a steady supply of excellent quality glass fibers being delivered to the many companies who use them to weave fabrics and make other products.

As you might assume from their lowly origins, the cost for the basic glass fiber is really quite reasonable. For many years it was about forty cents a pound, but with the increase in energy costs, of the past few years, it is now more than a dollar a pound for the lower cost yarns. This is still not very high when you consider how much fiber you can get out of a pound of glass. D-1800's yarn, for example, yields more than 104 miles of material in a single pound!

CARBON FIBER

Carbon fiber is probably the most commonly mentioned of the newer composite materials, and, like glass fiber, it is another material of rather lowly origins. Carbon fibers are usually made by oxidizing, carbonizing, and then graphitizing a yarn, in three separate but continuous operations. These operations are typically performed on a multi-filament strand of polyacrylonitrile fiber (PAN), which is provided in a form very much the same as that used by the weavers of synthetic fabrics. When the yarn is intended to be used as a precursor for carbon fiber, however, it is very specially controlled. This is because of the fact that the quality and nature of the original PAN fiber has a very strong influence in determining exactly what structural properties the resulting carbon fiber will have after it is manufactured.

In addition to being manufactured from rayon or PAN, carbon fiber is also produced from the pitch fibers obtained directly from petroleum pitch or coal tar. Theoretically, these pitch fibers provide a much lower cost method of making the final graphitized material, but so far this has not proven to be the case in actual production.

All of the carbon and graphite materials are much newer than the glass fibers. The first commercial carbon fiber was developed in the early 1960's, simultaneously but separately and independently, in both England and Japan. Domestic producers in the United States began as subsequent licensees of one of these pioneers. The technology has been advancing rapidly ever since.

Current practice in finishing nearly all carbon fiber requires a surface oxidation treatment, which makes the surface achieve a stronger bond to the matrix resin. This treatment is immediately followed by application of a sizing material, which acts as a primer, or pre-treatment, to further enhance the subsequent attachment of the matrix resin, as well as making handling much easier. Although there are as many different methods to accomplish these steps as there are fiber producers, nearly all of the fibers end up coated with a very thin layer of epoxy or similar resin. More oxidation usually provides a better bond to the matrix resin, but may degrade some of the other properties, while more or less sizing usually results in a change in the handling qualities of the fiber during subsequent processing. Typically, sizing is applied in an amount equal to one or two percent of the weight of the fiber, the heavier weights being used for weaving and the lighter used by prepreggers for making unidirectional prepreg.

TWO DIFFERENT FIBERS: PAN AND PITCH

Some significant differences exist between PAN and pitch fiber. PAN fiber is often preferred because, in a number of specific fibers, it has both higher tensile strength and higher compressive strength. Toray, for example, now offers T1000, a fiber having a tensile strength approaching one million psi, as well as several other types having strengths less than this, but higher than the usual 550 to 600 ksi of most PAN fibers. Pitch fibers, in contrast, are available only in strengths up to about 600 ksi. In addition, the compressive strength of pitch fiber tends to be somewhat lower than the compressive strengths available in comparable PAN fibers.

Most PAN producers also offer fiber having a modulus higher than the original 33 million psi (usually called, "msi"), such as 42, 45, 50 or 55 msi. Pitch fibers, however, are offered at modulus figures ranging from 5 msi to as high as 130 msi. The low modulus pitch fibers are used extensively in the production of carbon brake discs for aircraft and high performance cars, while the high modulus pitch fibers, typically in the range of 75 to 130 msi, are used mostly in spacecraft structures. Virtually no pitch fibers of intermediate modulus are used commercially. They are almost always in either the very low range of 5 to 15 msi range, or higher than 75 msi.

Filament diameter is also usually different. Most PAN fiber comes in a filament diameter of 5 to 7 microns, while pitch fiber is usually a little larger, from about 6 to 10 microns. (A micron is one millionth of a meter, or about 0.000039 inches.) Since a ten-micron fiber will provide about twice as much material in a strand of 3,000 filaments as will a 3,000 filament strand of seven-micron fiber, careful attention must be paid to this detail.

PAN fiber has a lower density than pitch fiber, about 1.75 to 2.0 gm/cc, as compared to 2.14 to 2.20 for pitch fiber. This difference can sometimes make direct comparisons confusing. Since the reason for the original choice of carbon fiber is usually weight saving, careful attention must be paid to the true density of the fiber you finally choose.

Other unique attributes of pitch fiber include rather high conductivity of both electrical current and heat. Values vary with both modulus and crystal size, so it is necessary to determine the exact value for the specific fiber under consideration.

One characteristic shared by both materials, however, is their sensitivity to the exact fiber alignment, fiber straightness, and surface defects in the final part. As a consequence, home projects which cannot insure low resin content, perfect fiber orientation and absence of waviness in the fiber often result in a far lower strength than that which was expected. The difference in the strength of a typical pultruded piece of "Spartuf" laminate or "Graphlite" rod, and that of a wet laid up unidirectional laminate of the same dimensions can amount to a factor of 5 or more!!

A peculiar thing about the carbon fiber business is the reference to the product as either carbon or graphite, almost interchangeably. This stems from a situation of long standing, in which the scientists involved could not seem to agree on exactly what it was that they were producing, some calling it, "carbon fiber", while other equally qualified authorities termed it, "graphite fiber." The matter still has not been entirely settled, and both terms are commonly used.

Each fiber maker produces a number of different fibers, the total of which make up his product line. This lineup of products may, or may not, be closely similar to that of any other manufacturer.

As a consequence, any product substitution by a user of these fibers will require careful testing and evaluation. This is a quite different situation from that which exists among glass fiber manufacturers, where each manufacturer produces equivalent products. This situation makes it quite easy to switch from one source of glass fiber to another without compromising structural performance, while such a switch from one carbon fiber to another may turn out to be a disaster!

Because of this situation you must, when using any carbon fiber material, specify both the manufacturer and the exact yarn made by that manufacturer. Since there are so many manufacturers, each with five, to as many as twenty or more different fibers in current production, the resulting confusion can become quite serious.

A summary of the "impregnated strand" strengths typical of a few of the many carbon fiber types currently available is shown in Appendix 5. Be careful in using these figures, as they represent the fiber only, before it has been cross-plied, woven or diluted with a substantial amount of matrix resin.

PAN BASED CARBON FIBER

In spite of the continuing mergers and bankruptcies, the business of providing carbon fibers has been so attractive to so many companies that there are still more than 15 manufacturers (including those of both pitch and PAN), all competing vigorously with each other. As of this writing, these include such firms as Amoco Performance Products, Hexcel, Aldila, Zoltek, Nippon Graphite Fiber, British Petroleum, Tofay, Grafil, Mitsubishi Rayon, Toho Rayon, Mitsubishi Chemical, Akzo-Fortifil, and others. Plant capacity of these manufacturers was far higher than current output for a number of years. However, in 1995 a sharp increase in usage resulted in a severe shortage of fiber in 1996 and most of 1997. At this point most of the fiber manufacturers were able to increase plant capacity, until, in late 1997, the balance began to recover. The industry capacity nearly doubled again, and has increased even more since. Now in 2005 the balance between capacity and needs still teeters back and forth but users are generally able to find suppliers of the material needed and new applications continually spring up everywhere.

The price of PAN based carbon fibers is another interesting subject. When the material was first introduced into the commercial market in golf shafts, it was quite common for a single golf club with a carbon fiber shaft to sell at retail for $500, or even more. Now the carbon fiber that goes into these products can be purchased on the open market for as little as $10 per pound, down from about $50 per pound just a few years earlier. The cost of the shaft alone, without head or grip, is now often less than $15.

The users of carbon fiber golf shafts have thus seen a tremendous benefit from the continuing price reductions of the fiber. The current consumption of one of the largest shaft manufacturers, Aldila, is said to exceed 10,000 pounds per week. Aldila now produces some of their own fiber.

The lower price for carbon fibers has caused their rapid growth into many other areas, such as tennis rackets and fishing poles, and includes the entire civil and military aerospace industry, as well as high performance power and sail boats and the chassis of most "Indy" type racing cars. More recently, this fiber has emerged as a leading candidate for use in strengthening bridge and highway structures and we have seen the emergence of an entirely new market of rather substantial size.

Carbon fiber has also found its way into automobile engines and bodies, where it occasionally performs a spectacular job at rather reasonable cost. One example is the use of carbon fiber in the main spring of a General Motors truck, resulting in a weight reduction of more than 80%. This astonishing weight saving is in addition to the dramatically longer service life provided by the carbon fiber part. Another example is the use of carbon fiber in the masts and spars of nearly all of the larger racing yachts, where they have offered substantial reduction in topside weight and improved ease of handling, often by a single crewman.

PITCH BASED CARBON FIBER

Adding to the confusion in the carbon fiber market is the emergence of the pitch-based carbon fiber materials as a specific solution to several of the problems inherent in aircraft and spacecraft structures. These structures are nearly always extremely weight sensitive, since the cost of putting the vehicle into orbit is of the order of $10,000 to $20,000 per pound of vehicle weight. Obviously the weight which can be saved by the substitution of carbon fiber for aluminum or beryllium can represent a real bargain.

In addition to this simple weight-saving goal, however, many space applications urgently require that some portions of the structure be dimensionally stable under wide changes in temperature. Such parts as antenna reflectors, microwave feed horns and their mounting structures, microwave plumbing, and multiplexer cavities, can seriously degrade performance of the vehicle mission if even a few micro-inches of movement occurs under temperature changes caused by either alternate exposure of one side or the other to the sun, or to heat generated by electrical units being operated.

ZERO THERMAL EXPANSION

Several specific advantages are offered by the pitch based carbon fibers. Their inherent capability of being designed and constructed to exhibit a "zero coefficient of thermal expansion", or CTE = 0, is one of the most important. This is made possible by the fact that these fibers all have a negative CTE in the fiber direction, making the fiber grow shorter with an increase in temperature, while the cross-fiber direction, which is dominated by the matrix resin, grows longer. In addition, the exact value of the CTE becomes more negative as the modulus of the fiber increases, allowing the designer to pick a particular value of CTE for his design, and produce a laminated structure which quite closely conforms to the chosen value in all directions. The value chosen is often zero.

Although these calculations make such a design very tricky, virtually all of the space vehicle manufacturers are now employing this method in the design of every new spacecraft. As a result, the newer designs continue to be much more efficient in their ability to transmit microwaves, and are able to use less power, while providing stronger, or a larger number of simultaneous transmissions.

HIGH STIFFNESS

In addition to being able to design to a zero CTE, the designers also have a broad choice of modulus for these fibers. The usual grades of PAN fibers may have a modulus of 33 (the most common variety), 42, 50, or even 60 msi and a little higher. However the commonly chosen pitch fibers may show modulus figures of 75, 105, 114, 120, or 130 msi! This is particularly impressive when one considers that the stiffness of steel is only 29 msi. The stiffness of the laminated structure is not as high as the stiffness of the fiber, which is the number always quoted by the fiber producer. However, even when the performance is reduced by the need to have fibers running in several directions and to be embedded in a matrix resin, the finished structure can exhibit surprisingly high stiffness. This stiffness can sometimes be as high as two to three times the stiffness of an equivalent steel structure, while weighing about one-fourth to one-half as much. Because space structures are operating completely out of the earth's gravitational field, very light structures made from pitch fiber components are possible, practical, and normally used.

HEAT TRANSMISSION

In addition to the remarkable properties noted above, pitch based carbon fiber can also transmit rather large amounts of heat along the length of the fiber. This capability is much higher than the similar capability of PAN based carbon fibers, and for the higher modulus pitch fibers it is even higher than aluminum, copper or silver! The approximate heat transfer capabilities of several fibers compared to copper and silver are shown in Fig. 1-3.

As a result of this characteristic, designers are beginning to use pitch-based carbon fiber as a primary method of extracting heat from densely packed electrical assemblies, such as those used in military communications equipment. It appears that the most favored approach to this problem is the use of a layer of pitch fibers within an otherwise normal circuit board, with at least one edge of the board adjacent to a heat sink or cooling source.

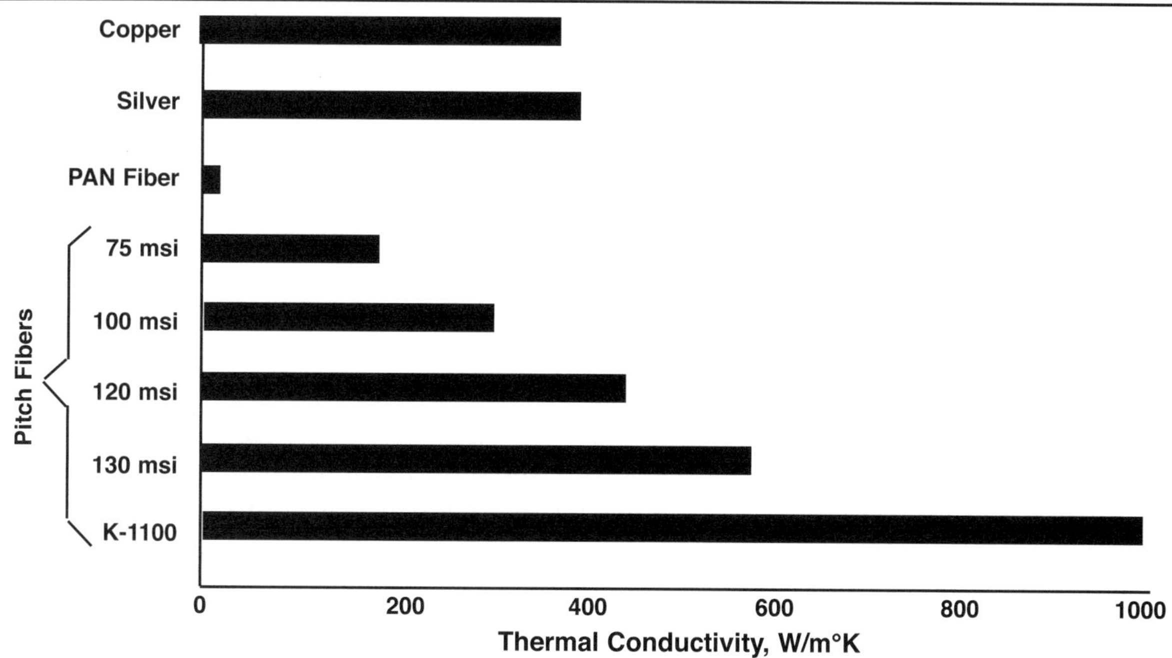

Figure 1-3. Thermal Conductivity of Several Materials

This attribute of the fiber also resulted in the use of pitch fiber honeycomb by the Boeing Company for some of the internal cowl panels on the nacelles of the 777, as well as newer model 737's. This particular honeycomb carries about the same amount of heat (or a little more) through the thickness of the sandwich panel as is carried by the same density of aluminum honeycomb, while offering complete freedom from the problem of galvanic corrosion between the previously used aluminum core and the carbon fiber facings of these panels. Development continues on this application of pitch based carbon fiber materials, and many more such applications will probably be seen over the next few years.

In addition to a brief summary of mechanical properties of several carbon fibers, Appendix 5 also presents information on a number of carbon fiber fabrics, woven from both PAN and pitch based fiber. Availability of any of these fibers will depend on the fabrics currently in use by the larger aircraft and spacecraft makers. Since both PAN and pitch fiber fabrics are in current use, most weavers will have at least a small selection of fabric materials available to casual buyers. Fabrics in the range of 75 grams per sq. meter to more than 300 grams per sq. meter (2.2 to 9.0 oz. per sq. yd.) are usually available.

THE NEW TOUGH PITCH FIBERS

Several years ago, Nippon Graphite Fiber Company introduced a new, and quite different class of low-modulus, pitch-based carbon fiber materials. The primary difference between these and earlier pitch fibers is the remarkable ability of the new fibers to absorb energy. This difference shows up as improved impact toughness in structures manufactured using small amounts of these new materials as a surface ply or plies. The most remarkable of the new fibers is called XN-05, which has a modulus nearly as low as glass fiber and a strain-to-failure of nearly 3%.

A flexure test specimen made from Toray T700, one of the very best of the PAN fibers, showed an improvement of more than 100% in fracture initiation energy when several surface plies on one side were replaced with XN-05! Design details necessary to reach such notable improvements can be quite tricky, and should be discussed with the fiber manufacturer if such an advantage seems desirable for your structure.

KEVLAR FIBERS

Another new fiber which has gained broad acceptance is Kevlar. This Dupont material has many of the attributes of carbon fiber because of its high strength and modulus. It also has a unique property to a degree which neither carbon nor glass fiber seems to have—toughness.

One form of the material, Kevlar 29, is used in bullet-proof vests worn by most of the police officers in the United States. It is also used in the new U. S. Army helmets, as well as the ballistic armor in ships and various other vehicles used by our armed forces. Kevlar 29 is also extremely light—it weighs about half as much as aluminum.

The Kevlar which usually finds its way into load-bearing structures is Kevlar 49, and more recently, Kevlar 149. These fibers have equal strength, but have a much higher modulus than Kevlar 29. Unlike carbon fibers, Kevlar 29, 49 and 149 do not conduct electricity, nor are they electrically opaque to radio waves—two of the problems with carbon fiber. They also have low compressive strength and modulus compared to their tensile properties. This problem is somewhat the same as encountered in carbon fiber, but is a little more pronounced in the Kevlars. Kevlar acts much like glass fiber, but even better, for structures used to transmit microwave radiation, such as radomes and antenna windows. The diameters of Kevlar fibers and the methods used in handling them are similar to both carbon and glass fibers, but the form in which the material is available is different.

Kevlar is produced in only one fiber size. These fibers are used to produce yarns of various weights, or "denier" grades, which range from 195 denier (a bundle with 134 individual fibers) to 7100 denier (with proportionately more individual fibers). Dupont is constantly revising the array of products they offer and can provide deniers of up to over 20,000 for very heavy applications.

The use of the term, denier, is a little confusing because in the composite structures business, nobody uses the denier system except Dupont. If you buy yarns or rovings of graphite you will be offered a 1K, 6K, or 12K strand. The "K" refers to the number of filaments in a strand. A 12K yarn has 12,000 filaments in a single strand. An 880 denier yarn of Kevlar has 267 filaments which makes it a little less than one-third the size of a 1K yarn of carbon fiber. The term "denier," is defined as the weight in grams of 9000 meters of the strand. Although one is tempted to attribute the logic behind this definition to the fact that, "the system was developed by the French," the real origin lies in the fact that 9000 meters is the approximate length of a single strand of silk spun by a silk worm.

Compared to other products used in similar structures, Kevlar is actually a very good buy. The yarn has been offered to nearly all of the weavers and many different styles of fabric are available from at least 10 manufacturers, who produce fiberglass fabric and carbon fiber fabric. The weavers compete vigorously with each other and as a result the woven fabrics are readily available and reasonable in price.

NEXTEL FIBERS

Nextel is an interesting new fiber from 3M Company. It has a strength a little less than glass, and has about the same density. Its remarkable attribute is its retention of properties at temperatures in the range of 2000 degrees F, combined with a very low heat transmission rate. As a result, it appears to be the best candidate—by far—for firewall, and other fire resistant structures. On a weight basis, the material dramatically outperforms stainless steel in this application! Fabrics intended for this end use are produced by many weavers. Data on the material can be obtained from either the weaver, or directly from 3M.

SPECTRA FIBER

During the past several years, a new fiber has been commercialized by Allied Signal Corporation. It is called by the trade name, "Spectra", and is becoming one of the standard solutions for the same sort of applications as the Kevlar fibers. There are several points of difference worth noting. First, the surface of Spectra fiber is so difficult to bond to, that few laminate applications are possible.

The second point of note concerning Spectra fiber is its very low specific gravity. It is the only one of the structural fibers to float on water! This attribute makes the fiber a popular choice for body armor, helmets, and other such weight sensitive products. Also, its remarkable resistance to degradation under salt water exposure has resulted in promising applications to marine rope and hawsers, mooring lines and so forth.

Although it is still too early in the life of this fiber to be certain of its future position among the structural fibers, it appears to be about to displace at least a portion of the substantial markets now held by the Kevlar fibers.

THE FAR-OUT FIBERS

When we talk about advanced composite fibers, we usually use the word "advanced" to indicate that the materials have been developed recently. We often include E-glass, even though the product is nearly 60 years old, because its performance is so outstanding when properly used.

There are a number of fibers, however, which are considerably more advanced than even the Kevlar and carbon fiber materials. These are the borons, the carbide whiskers, the silicon carbides, carbon nanotubes and nanofibers, as well as many other unusual, refractory fibers. Some versions of these materials have been used not only as continuous fiber reinforcements, but as short fibers and as "chips", for use in strengthening cast aluminum alloys. These materials not only have extremely high strength and modulus, but many can be used structurally at temperatures in the incandescent levels, in the range of 1500 to 2500 degrees F.

Consequently they are of great interest to the Air Force Materials Laboratory, where more than 10 such fibers are under study. Some of these may turn out to be commercially significant, and will show up in commercial or amateur-built structures in future years, just as S-glass, Kevlar and carbon fibers do right now. For the present, however, most of the technical information is classified and the cost of available materials seems astronomical, often over $1,000 per pound, about the same as carbon fiber cost 40 years ago.

An excellent review of 20 or more of these new fibers is contained in pages 152 through 167, Vol. 2, *International Encyclopedia of Composites,* edited by Stuart M. Lee and published in 1990 by VCH Publishers, Inc. The section is *Fiber Reinforcement, High Performance.*

THE FIBER/RESIN INTERFACE

Basically, the resin matrix is the key to the whole operation of producing composite structures. It was noted earlier that the resin matrix is the mass in which the fibers exist, but the resin does much more than just contain the fibers. Its primary job is to carry the load from one fiber to the next, and from the bundles of fibers or groups of reinforcements into an adjacent structure which may either be embedded in the composite during manufacture, or adhesively bonded to it at a later stage. The resin material thus distributes and transfers the load within the structure so that each reinforcing fiber carries a proportionate share of the total load.

A basic question is that of achieving a strong, permanent and reliable bond between the fiber and resin, so that the load can get out of one fiber and into the next—even after aging or long exposure to the sun. Because it is so important, this fiber/resin interface has been the subject of a great deal of work within the composite industry over the last fifty years. In the case of E-glass fibers, a large body of literature and a well developed technology are in current use by most manufacturers and users.

The primary problem is that most resins which appear to give a good bond to the untreated fiber will allow that bond to deteriorate over time—sometimes during a period as short as a few weeks. For example, if untreated glass fibers are bonded with general purpose polyester resin (like that used in most recreational boats and many modern bathroom fixtures), the bond will appear to be quite satisfactory the day it is made, but two or three weeks later, will show a drop in strength of 5% or 10%. Three or four months later this drop could be 30%, and a year later, as much as 70%! The situation which leads to this is that the interface between the resin and the fiber is deteriorating and the strength of the original bond has nearly disappeared. With glass, this deterioration usually involves the addition of a water molecule to the surface of the glass, forming a new substance which is not a very good structural material.

To prevent this deterioration, and improve the "wetting" of the glass fibers by the resin, it is a common practice to treat the surface of the glass with a coupling agent before using the glass in a composite structure. This treated fiberglass surface will make a permanent and lasting bond to the resin. All of the weavers and prepreggers will apply one of any number of such treatments, usually called "finishes", to the cloth, using the finish they deem most appropriate to the end use of the cloth.

For example, one of the oldest and still a most commonly used finish for E-glass is called "Volan". This material is actually a methacrylic chromic chloride complex and is applied directly to the clean glass fibers. The Volan finish is a good one, and this finish, or one of the many newer variations of it, have been widely used. It gives a good bond to both epoxy and polyester resins and also to the newer vinylesters and polycyanates. As a result, this class of finish is commonly used for glass fabrics intended to be used in both boats and homebuilt aircraft.

If all E-glass fabrics had one of these finishes, most people using it in hand layups would not have much of a problem. Unfortunately, a lot of fabrics are intended for totally different end uses, and have one of the many other finishes, which may be very unsuitable, as noted below.

A different possible problem is that most fabrics in the "as woven" condition have a starch and oil "binder," or "sizing," on them. This is a lubricant which enables the fragile glass yarns to get through the weaving operation without tearing and destroying the structural integrity of the fabric. Fabrics with the starch and oil still in place are called "greige goods", or "loom-state".

These materials on the surface of the fibers can cause serious problems if the fabric is used in a reinforced plastic or composite structure. Water goes right down the interface between the resin and the yarn because of the hydrophilic (water-loving) nature of the glass. Therefore, it is common practice to burn off this binder after the fabric is woven, giving it a "heat cleaned" surface. The surface finish is then applied to the heat cleaned cloth to make it suitable for the intended final use.

The problem doesn't end here, however, because many other surface finishes are applied for totally different end uses. "Release fabric," for example, may be the very same glass cloth you are using to make a structural laminate, but with a finish intended to prevent adhesion to resins. Other finishes are applied for the purpose of preventing the glass from chafing itself when used for fume filtration in a bag house. There are more than a hundred different, commonly used fiberglass finishes which may appear in woven glass cloth. All but ten or twenty of them will do just the wrong thing to a composite structure.

It is, therefore, very important that every user of structural glass fiber knows:

1. where his fiberglass cloth came from.
2. that it has been properly treated with a finish that will deliver the structural result and the durability he is looking for.
3. that it still has the finish in place, and that it has not been contaminated by water, smoke, coffee spills, grease, oil, etc.

A standard precaution is to avoid surplus glass fabric for use in composite structures, because you never know where it came from. Since most finishes on glass fabric tend to deteriorate somewhat with age, it is a good idea to buy your glass in smaller quantities—no more than you will use up within a year or two. (Use any suspect material in non-critical parts.)

Although Volan is the only surface finish that has been mentioned, there are probably ten or twenty more very good ones which are used by the weavers and prepreggers.

The important thing is to be sure that one of the appropriate finishes was actually applied to the fabric you intend to use, and that you use a resin system appropriate to the finish.

Figure 1-4 lists the different finishes available from one weaver, Hexcel, several years ago. It can be noted that the list includes heat-cleaned glass, Volan, and many other finishes, each of which is intended to perform a specific task under very specific conditions.

With the fiber-resin interface on Kevlar, we have a different situation. The material cannot be heat cleaned because it is an organic substance and cannot withstand the 700

Trevarno F12
Removal of all the original binder through oven burning, leaving virtually pure glass having a residue of approximately 0.1%. Not recommended for fabrics less than .04" thick. All fabarics which are subsequently finished with coupling agents go through the heat cleaning process. Resins recommended: Silicones.

Trevarno F13 (neutral pH)
Heat cleaned fabrics are washed in demineralized water providing a neutral pH. Resins recommended: Silicones.

Trevarno F16 (Volan A)
jF12 heat cleaned fabric is saturated in a methacrylate chromic chloride solution, cured and then washed to remove any soluble salts. The chrome content of the finished fabric is between 0.03% and 0.06%. Polyester laminates using Volan A finished fabrics meet the requirements of Federal Specifications L-P-383 and Mil-C-9084. Resins recommended: Polyesters, Phenolics and Epoxies.

Trevarno F39 (A172 Union Carbide)
A vinyl-silane is applied to F12 heat cleaned fabric. Laminates using F39 finished fabrics meet the requirements of Federal Specifications L-P-383 and Mil-C-9084. Resins recommended: Polyesters.

Trevarno F45 (A1100 Union Carbide)
F12 heat cleaned fabric is completely saturated in an amino-type silane solution and then cured. Pickup is approximately 1.5%. Laminates using fabrics having F45 finish meet the requirements of Federal Specifications Mil-R-9299 and Mil-R-9300. Resins recommended: High-temperature Phenolics, Epoxies and Melamines.

Trevarno F40 (A1100 Soft — Union Carbide)
A reactive amino silane solution is applied to F12 heat cleaned fabrics and then dried. Fabarics having an F40 finish have greater drapability and softer hand than those finished in F45. Laminates made with fabrics having an F40 finish meet the requirements of Federal Specification Mil-R-9299 and Mil-R-9300. Resins recommended: High temperature Phenolics, Epoxies and Silane Modified Phenolics.

Trevarno F43 (A 174 Union Carbide)
F-12 heat cleaned fabric is saturated in a mixture of acrylate modified trimethoxysilane in a water solution and then cured. Fabric pickup is 0.15% to 0.20%. Resins recommended: Polyesters.

Trevarno F48 (Z6020 Dow Corning
F12 heat cleaned fabric is completely saturated in an amino type silane solution and then cured. Fabric pickup is approximately 1.5%. Resins recommended: Phenolics, Epoxies and Melamines.

Trevarno F49 (Garan)
F12 heat cleaned fabric is saturated with a silane type coupling agent which leaves a vinyl-triethoxy on the surface of the glass. Trevarno F49 gives outstanding wet and dry strength properties. Polyester laminates using F49 finish meet the requirements of Federal Specifications L-P-383 and Mil-C-9084. Resins recommended: Polyesters and Epoxies.

Trevarno F5 (weave set sizing)
Sizing applied to cloth styles used in liquid filtration.

Trevarno F6 (weave set sizing)
Phenolic sizing applied to cloth styles used in liquid filtration.

Trevarno F8 (weave set sizing)
Finish applied to greage goods for pre-primed Navy Board and similar applications.

Trevarno F21 (weave set sizing)
White sizing applied directly to greige goods for use as a pipe wrapping. This finish carries underwriter's label.

Trevarno F51
Blue colored PVA applied to various cloth styles to be used as release fabrics. Excellent drapability and most economical.

Trevarno F53
Special silicone finish applied to special weave glass fabrics designed for fume filtration.

Trevarno F54
Pink colored silicone finish applied to various cloth styles to be used as release fabrics. Excellent drapability and most economical.

Trevarno F58
Tan colored cured silicone rubber polymer applied to various cloth styles to be used as release fabrics. Recommended where parts have secondary operations.

Trevarno F68
Special graphite-silicone teflon finish applied to specail weave fabrics designed for fume filtration.

Figure 1-4. The Various Finishes Used by One Weaver for his Line of E-Glass Fabrics

to 800 degree temperatures which are used to burn off the sizing and binder from ordinary E- and S-glass. Consequently, the sizing and binders are washed off Kevlar fabrics in a washing machine, a process called "scouring". A special, water-soluble binder is used to permit this step.

If you use Kevlar fabrics as a part of a composite structure, you must be certain that they have been properly scoured before you start applying them. It's easy to tell if they have been scoured, because if they have not, the cloth will be very difficult to wet out and simply won't look right when the part is finished.

Although you might think there would be some chemical surface materials to treat Kevlar yarn to improve the bond to the resin, so far a completely satisfactory one has not been developed. Dupont continues to work on this problem and it is expected that, within the next few years, Kevlar will be more readily bondable to a larger array of resin systems. At present it should be used only with epoxies or vinylesters. The polyesters simply don't bond to Kevlar at all well, yielding rather low strength values in the finished structure.

In spite of the lack of a surface treatment for Kevlar, several specific things should be noted in this regard. First, the tendency of the molecule to pick up water, up to seven or more percent by weight, means that most resins will bond poorly to the fiber if it is not first thoroughly dried. This can be accomplished by storing the roll of fabric or fiber in a heated storeroom, such as a closet with a 100 watt light bulb left on continuously. Be sure to give the material enough time to lose all the water it has picked up. Weighing the roll is a good way to tell if it is still losing weight or has stabilized. It has also been found by some users, especially prepreggers, that treatment of the material by exposure in an electric Plasma Chamber often solves interlaminar shear problems. And, of course, always be certain that the material has been scoured before you receive it. If it has not been so treated, almost no resin will wet it at all.

There is also a difference in the level of strength developed by the various formulations of resins which do bond well to the fiber. If you are going to make a Kevlar and resin composite, it's a good idea to test the system that you propose to use, and to test the mechanical properties of the finished composite before you commit a large scale construction project. In some cases, the resin manufacturer may have already run these tests and can furnish data on the combination you propose. However, do not assume this unless you actually verify it.

Carbon fiber materials cannot be either heat cleaned or scoured and, therefore, cannot be given the same lubrication treatment that E-glass fibers are given for weaving. As a result, the lubrication of these yarns is always accomplished by putting a little bit of epoxy resin on the surface of the carbon fibers. This occurs after they are drawn from the oven and before they are wound on the bobbins.

This bit of resin, which may range from less than 0.5% to more than 2% by weight, is then joined to the resin that goes into the final product. You must be careful that the resin system you choose when using carbon fiber is compatible with the fiber finish already in place on the material. For example, if you use a polyester resin system with fibers that have an epoxy resin lubricant, it's quite possible that you will get very bad structural results. Even some of the epoxy systems can give similarly disappointing structural values in cases where the matrix resin is not compatible with the finish.

Spectra 900, or Spectra 1000, is somewhat similar to Kevlar in that it has turned out to be very difficult to bond to, and as of this writing, no satisfactory finish is available to enhance the bond. The only glimmer of light is the use of plasma treating on the surface. It appears that some resin systems may bond well to this treated surface, and finally provide a laminate which has the strength suggested by the fiber properties.

Zylon fiber is another of the very new and promising materials beginning to appear. It is available as fiber wound on a bobbin (to the weavers), and as woven fabric. It is somewhat like Kevlar, in that it has a light brown color, and resists bonding to most resins. It has been found that plasma treatment can offer reasonable bond strength to Epoxy resins, but much remains to be learned before the material is suited for inclusion in a home project. One caution is the fact that the material deteriorates rapidly in sunlight, substantially worse than unprotected Kevlar. At the present time, it certainly appears that much more must be learned about the material before it can be used by anyone on a confident and reliable basis. The very high modulus and strength figures of test samples make it a very attractive candidate, however.

CHAPTER 2

FABRICS

WEAVING THE FIBERS

Let's look at how the weaving process affects the performance of materials that we might put into a composite structure. Typically, the weaving done with any of these materials is a direct descendant of the same weaver's art that has been practiced for thousands of years all over the world. In fact, some of the looms used, particularly for narrow tapes, are very old designs with the patents long run out. Some of the machines themselves are forty to eighty years old!

For the newer and lighter fabrics, however, the looms quite typically will be of more recent design. Some of them are quite capable of very high speed operation. In the weaving industry, while the level of technology is high, the level of cost addition to the products is usually quite low. In other words, with woven fabrics you usually get a lot for your money.

For example, Kevlar 49, which sells for $20 to $40 per pound as a yarn, is made into a fabric that weighs 15 ounces per yard and is currently available in relatively small quantities for as little as $25 per yard. With carbon fiber materials this is not so apparent as carbon fiber is even more difficult to weave than Kevlar, resulting in lower weaving speeds, higher costs and higher losses during the weaving process. All of these contribute to higher prices for the fabric. Even so, every time there is a glut of carbon fiber in the marketplace such as that which occurred during the late 1980's and early 1990's, the cost of the most common fabrics has dropped to levels as low as $15 to $30 per yard.

With glass fiber materials there is so much competition and the art is so well understood by those who practice it, that the woven raw goods are available with even lower markups than Kevlar fabrics. As a result, glass cloth continues to be one of the all-time best buys in the field of structural products, just as it has for the past thirty or forty years.

One weaver's 1989 price list, for example, shows a style 7781 fabric (about 0.009 inches thick per ply and weighing about 9 ounces per sq. yd.) in a fifty-inch width, available for $2.92 per yard for a 125 yard roll. Even very light fabrics such as style 116 (about .0035 inches per ply) are available for about the same price. This is rather startling when one considers the degree of skill and art that is required to produce the material involved.

One of the problems that the average aircraft or boat builder faces is that of inability to deal directly with the weaver, who does not usually want to be bothered with small orders of less than a full roll. You can expect to pay a small premium when you buy from one of the normal suppliers who will inventory the material, and then recut and package it for reshipment to individuals.

Even then, the prices are not all that high. As a general rule of thumb with glass cloth, the yarn will cost from $0.85 to $1.50 per pound, and the woven fabric will be about double that. If the cloth is carried in inventory by a supplier for sale to individuals, its price will approximately be doubled again. Fortunately, these multiples do not usually apply to carbon fabric and Kevlar fabric, because they simply are not marked up the same way glass cloth is.

Glass yarn is available in a broad choice of sizes and descriptions, and the number and descriptions of the fabric weave styles available far outnumber the types of yarn. Just to list the types of glass fabric in current production in the U. S. would probably take about a hundred pages. However, all of these styles are not produced for composite structures. Composite structures use a relatively small list of fabric styles and a relatively small number of yarns, although in fiber glass this can include more than 24 different yarn styles.

There are several facts of life involved in the use of fibers in composite structures which should be kept in mind. Number one is that the fiber which carries the load can carry it best if it is absolutely straight. If a woven fabric is to be used, the less crimping of the fiber involved, the better. Thus it is normally seen as an improvement in the structure when cloth weaves go towards the eight-harness satin weave style, in which a single strand of yarn goes over seven strands and under one strand to make woven cloth. The satin configuration is quite common in composite structure and several different versions are used in very large quantity.

Figure 2-1 shows details of construction of four common weave styles. In this example, the satin weave shown is a "4-harness satin", or a twill. An 8-harness satin would have the yarn tuck under every eighth crossing yarn, instead of every fourth crossing yarn, as shown in the example.

There is also an improvement in the strength of the fabric when it is made from an untwisted, single yarn rather than one of the same total weight that is twisted and plied. Both styles are available in the same weave, and the commonly used 7781 fabric also shows up as a 181, and a 1581. The different numbers represent an eight-harness satin of about nine ounces per square yard with exactly the same thread count and weight. However, the 181 has a twist to the yarn which is more exaggerated than that of the 1581, and each yarn of the 181 is formed from three strands, each of which is a twisted pair, for a total of <u>six</u> strands of fiber bundles, each one of which has its own twist.

This particular yarn, used in style 181 glass fabric, appears as the example at the bottom of Fig 1-1, in Chapter 1. The 7781 is made from a single, untwisted strand, having the same total number of filaments as the plied yarn of Figure 1-1. The difference in strength shown by these fabrics in a composite structure is about 5 to 10%, favoring the 7781. And, because 7781 is also slightly lower in cost, it has largely replaced the older versions of this fabric.

Another important consideration in the use of fibers to carry loads is that the fibers need to be placed in the same direction as that in which the load must be carried.

If you are to build a spar cap for a conventional wing structure in which the spar carries all of the wing bending load, then the spar cap's fibers need to go in the direction of the wing's span. This is why we see unidirectional cloth, or roving right off the spool, commonly used in spar caps. The same is obviously true in the case of a mast or spar for a boat.

Fibers are available as either fine yarn, discussed above, or as rovings, which are very thick yarns, usually wound on a fiber tube, or simply folded into a fiber drum. They are also used in the production of unidirectional tapes, in which the strands are stitched together so that you end up with a flat sheet made up mostly of yarns going in the same direction, with a few rows of stitching to hold them generally together. They can also be supplied as unidirectional woven fabrics, in which most of the yarn goes in one direction, with only a few yarns going in the cross direction. The ratio of strength in the 0 degree

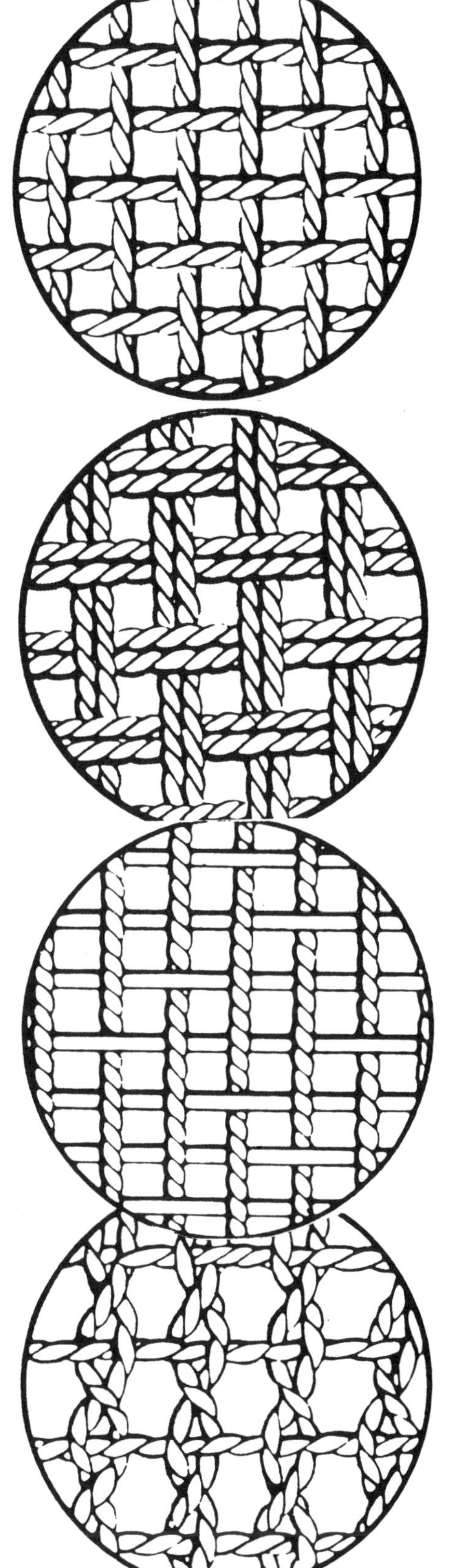

PLAIN WEAVE
The plain weave is the most simple weave pattern, with each strand going alternately over and under the crossing strands. The example shown uses a plied yarn made up of two smaller yarns twisted together. It also appears to be a balanced weave, with the same count and construction in both the warp and fill directions.

BASKET WEAVE
The basket weave uses two yarns next to each other, in place of a single yarn, to make a plain weave.

TWILL or SATIN WEAVE
In these weave styles, one yarn crosses over several yarns before going under a single yarn. If it rides over two, it is a Twill. If over three or four, a Crowfoot, and if over more than four, a Satin. It is a Five-harness Satin if the strand goes over four and under one, and an Eight-harness Satin, over seven and under one.

LENO WEAVE
In an open weave, with considerable space between yarns, this style is used to lock the yarn in position. This permits handling of the fabric with a minimum of displacement of the yarns.

Figure 2-1. Four Common Structural Styles

direction (the "warp", as compared to the 90 degree, or cross direction (the "fill", can be as high as fifty-to-one, and covers the complete range right down to one-to-one. Structural fabrics may be obtained in different weights, with thicknesses down to less than 0.001 inches. A glass cloth of style 104, for example, in a single ply of a composite structure has a thickness of 0.0012 inches, while a style 116 cloth has a thickness of 0.0035 inches, a style 7781 has a thickness of 0.009 inches and a style 7784 a thickness of 0.026 inches. Since there are many other standard fabrics in this range, the different weights available in standard fabrics turn out to be almost without limit. Some of these "standard" weave styles are listed in Appendix 2. On the high thickness end of these materials we can get "fine" fabrics in which the thickness of a single layer of fabric will come out at about 0.017 inches. If we consider the coarsely woven, heavy fabrics, such as woven roving, then this thickness can go up to 0.022 inches, 0.045 inches, and even higher.

All of the foregoing comments about the versatility and availability of woven fabrics apply only to glass fibers. There are not nearly so many choices available in Kevlar or carbon fiber fabric. However, with each passing year the weavers offer new fabric styles and it is now very seldom that a designer must have a new style woven in either carbon fiber or Kevlar.

A particular difference has developed between the usual glass fabric when compared to the usual carbon fiber fabric. The carbon fibers are much more sensitive to the "out of plane" condition caused by twisting of the fibers into a twisted yarn, and almost never are seen in any twisted configuration. The straight fibers shown in Figure 2-2 and on the cover show the parallel fibers nearly always seen in these fabrics.

SPECIAL WEAVES

If you decide that you really must have a fabric that is slightly different from anything currently available, the weavers will be glad to put one together for you. There are a

Courtesy Nippon Oil Co., Ltd.

Fig. 2-2. Parallel Fibers.

great many instances where a special fabric has been designed and woven for a special application. For example, Burt Rutan had two special fabrics, style 7715 and 7725, woven specifically for his VariEze/Long-EZ series of airplanes.

Special weaves are not any more costly to produce than standard fabrics. But, because there is only one application and one customer for the material, the weaver usually requires a rather large minimum order. In Rutan's case, he was required to order 50,000 yards to get his project under way. These particular fabrics are now currently available through most of the usual distributors, as they are used on many designs in addition to those from Rutan.

NON-WOVENS

An entirely separate class of fiber web reinforcing materials is the broad category of "non-wovens". These include both chopped fibers held together with a resin-soluble sizing, or binder, as well as continuous strand mat, and various unidirectional materials. The chopped and mat materials are much weaker than fabrics, even in the finished composite structure, and are not considered as being suitable for aircraft, high performance boats or racing car chassis applications. The unidirectionals, however, can be as strong as, and sometimes even stronger, than the best woven fabrics.

These materials are available as dry fibers or as pre-impregnated unidirectional tapes. The prepreg version is available from all the prepreggers listed in Appendix 1. The dry fiber multi-axial types are available from Hexcel (Knytex), Gulf Paper (COFAB), and several other manufacturers. They are available in 0/90 and +-45 degree biaxial orientations as well as 0/+-45 triaxial orientations.

Single layer unidirectional reinforcements are available from Aerotex, Anchor Reinforcements, and Hexcel. These materials can yield particularly efficient structures. Characterized by flatness, low crimp and high fiber volumes, unidirectionals allow the designer and builder the ability to apply fiber in precisely the same direction as that of the loads. They are often used in the fabrication of wing spar caps.

HYBRIDS

One of the interesting aspects of the different fibers used in composites is that they can be blended into a combination of fibers and then woven into fabrics. The properties of the resulting composite are quite different from properties of a composite made of only one of the fiber types.

For example, if we use S-glass to make a composite it ends up with a very specific set of mechanical properties which are quite different than those that would result if carbon fiber were used to make a structure of the same size, weight, thickness, and fiber orientation. If we use a yarn which is half carbon and half S-glass fiber, we end up with a composite structure that has a character all its own. It is possible to alter the performance of a composite a great deal just by mixing in a small amount of a different fiber.

A composite skin with S-glass as the reinforcement, exhibits a relatively high degree of toughness, but, when compared with a carbon fiber skin, the S-glass is not nearly as attractive in tensile strength. Carbon fiber skin is rather brittle compared to the S-glass skin. If we make a yarn with 50% S-glass and 50% carbon fiber, we find that we get nearly the same strength as the carbon fiber skin and the same toughness as the S-glass skin. The surprising thing is that if, instead of using 50% carbon fiber, we use only 20% carbon fiber, we get virtually the same results.

Consequently we are now seeing combinations of S-glass with carbon fiber, S-glass with Kevlar, E-glass with Kevlar, carbon fiber and Kevlar, and even PAN-based carbon fiber with pitch-based carbon fiber. These hybrids are beginning to show up in production structures, such as the wing-body fairing on the Boeing 757 and 767, as well as the high thermal conductivity carbon fiber honeycomb used in the Boeing Model 777 nacelle structure. These include laminates made of several plies of carbon fiber with the surface ply made of Kevlar 49 to improve the overall toughness and abuse resistance of these relatively lightweight panels, as well as honeycomb made with hybrid PAN/pitch carbon fiber yarn.

One of the applications of hybridizing, which amateur aircraft builders are already beginning to use, concerns Kevlar. Unfortunately, Kevlar is very difficult to cut and sand in its finished, fully cured form. When you take sandpaper to it, the result is not a nice, smooth surface, but a mass of fuzz which will not accept paint properly. The resin matrix is removed by sanding but the Kevlar fibers are so tough that they are simply released from the resin and start to build up on the sanded surface until they become fuzzy, lumpy and out of control. Consequently, a finished molded surface of Kevlar must be used in its as-molded condition and not finish sanded in the way that a molded glass fiber surface can be.

The simple way in which this problem has been overcome is to make a multi-ply structure in which most of the thickness is comprised of Kevlar layers, but the final surface layer is a thin, woven glass cloth. This outer layer of glass cloth is there simply for the purpose of sanding. When the finished structure is dressed out to get the cosmetic appearance desired, the glass layer sands and handles in a very satisfactory manner. It also protects the underlying structure from being cut by the sandpaper and having its structural integrity degraded or destroyed.

The thickness of the glass cloth to be used on this final layer is a function of how much sanding must be done on the final product. If it is only to be slightly dressed, a style 108 or 112, or even something as light as a 104 fabric can be used, which results in a very small weight penalty. On the other hand, if a considerable amount of sanding will be needed in order to change the shape substantially, a heavier layer of cloth, such as six or ten ounce tooling fabric can be used.

THE IMPORTANCE OF THE `FINISH,' OR COUPLING AGENT, USED TO ENHANCE THE BOND OF THE RESIN TO THE FIBER

In early 1983 VariEze and Long-EZ builders were startled to read in Burt Rutan's "Canard Pusher" newsletter that some defective UNI fabric had turned up in finished parts. The specified material in the RAF plans was a style 7715 fabric, produced by Hexcel and distributed at that time by both Aircraft Spruce and Specialty, and Wicks. The problem centered on a closely similar fabric, which was also designated by its manufacturer, Burlington Industries Inc., as style 7715, but which yielded lower strength when tested in identical layups. A short time later the same warning appeared in the "Dragonflyer", a newsletter to Dragonfly builders. A recall of the questionable fabric was subsequently undertaken by Burlington.

At this point most industry people expected the entire matter to disappear. Instead, Burlington apparently decided that the market was big enough to make it worth fighting for and did some testing and further development in an effort to correct the acknowledged deficiencies in the fabric. In a letter that was then circulated by Alpha Plastics, Burlington stated that the deficiencies had not only been corrected, but that test results on the Burlington fabric were said to yield even higher figures than the Hexcel material!

The situation which this affair highlights brings to mind several thoughts, as follows:

1. The 7715 designation for the Rutan UNI fabric, and the 7725 designation for BID are numbers that were given to these fabrics by the Hexcel development laboratory in the mid 1970's, when they were first woven for the VariEze project. The fabrics themselves, however, are much the same as two fabrics that were already in use in Europe at that time. I have not been able to determine whether the Hexcel fabric is an exact duplication of the European material or simply a similar style suggested by the construction of the earlier product. This is really irrelevant, since all the subsequent Rutan developments, as well as most other glass-over-foam designs have since used the same material from Hexcel.

2. The product designation numbers, or "style" numbers used by Hexcel to describe the two fabrics are a part of a loosely integrated numbering system that is routinely, but not always, used by all of the weavers, as can be seen in Appendix 1. Had Hexcel known at the time it first used this style designation that the commercial volume of the materials would be either this large or this critical, it can be speculated that the company would have used a more unusual designation, which Burlington might have been less likely to adopt.

3. The practice of copying a weave style and using the same product designation as that used by the originator of the style is common in the weaving industry. It is probable that if Burlington had duplicated the laminate mechanical properties of the style 7715 fabric more closely, no question ever would have been raised by Rutan, and the homebuilders would now have two suppliers along with the highly competitive pricing common in the glass fabric industry. For aircraft structures, however, mechanical properties are paramount, and no uncontrolled materials should be accepted or used.

4. Just exactly what mechanical properties will be observed in a finished laminate will depend upon several factors. Of course, much will depend upon the strength of the fabric used, but the "finish", a chemical coupling agent used to improve the bond between the fiber and the resin, is also quite important, as are the matrix resin and the process by which the laminate is assembled. Burlington, at the time the problem was current, made no statement with regard to any change in yarn construction, yarn count, fiber size or weave style. The **only** change acknowledged was a change in the coupling agent, from its "1-627" to "1-617". This change was said to improve the laminate strength such that it is superior to laminates made from the Hexcel fabric. Unfortunately, the Burlington letter failed to present any data in support of its statement.

5. The coupling agents used by various weavers may be either one of the older treatments, such as "Volan", or one of the newer (usually proprietary and usually better) finishes that yield higher properties in the finished laminate. Hexcel has a proprietary numbering system that identifies each system as a "Trevarno F(XX)" finish on the fabric. The company felt that, in this situation at least, the finish being used on its 7715 material was so proprietary that it refused

to publicly identify it. This sort of protective stance on the part of a manufacturer makes it difficult to endorse their product, even though it might appear to be the only reliable choice.

6. The testing to be done on a glass fabric in order to be sure that it will produce the anticipated strength level in a finished laminate is often extensive, complex and costly.

   If you are a major airframe maker, you may choose to attack the problem head-on, making thousands of tests, and subsequently writing and enforcing both material specifications and process specifications. All of the technology needed to obtain design values for all of the composite materials that amateur projects use is well established, but the cost to accomplish the job is well beyond the financial capability of most smaller companies.

   Instead, designers of boats, racing cars and amateur-built aircraft use the simpler and lower-cost method of choosing established materials and processes which appear to be best suited for the job, and load test the finished structure with whatever degree of thoroughness is needed. The remarkable service record of some 2,000 VariLoops, VariEzes and Long-Ezes, as well of that of the thousands of small boats and hundreds of race car frames built and used speaks very positively as to the success of that approach.

   The ability to substitute materials under this system is near zero, however, unless the original structural development and test programs are entirely repeated for each instance of substitution. In other words, if you use a new resin system or a new fabric finish, you have a new and unknown structure at hand, one which must be subjected to the same scrutiny and evaluation as if it were a new, "from-scratch" design. Since this is a fairly well-established fact in the composite industry, it would appear that Burt Rutan's seeming stubbornness in refusing to accept substitutions of specified materials is very well-founded.

7. Neither Hexcel nor Burlington would normally conduct structural testing of laminated composite samples as a routine or continuing method of quality control. Instead, they usually verify fabric strength by tensile tests of dry fabric and finish quality by wet chemistry or surface analysis methods. When a need for an additional receiving inspection or quality control inspection of the fabric seems appropriate, the fabric buyer will usually use the same tests.

8. In the absence of a materials laboratory, it is still possible for a back yard project builder to compare the strength of different composite structures, given some low-cost equipment, the urge to test and a little knowledge. The subject of load testing your structure is covered in Chapter 12, and includes some examples of testing your own laminates.

CHAPTER 3

CORE MATERIALS

HOW CORE WORKS IN A STRUCTURAL SANDWICH

The structural sandwich is simply a particular arrangement of materials in which we build the composite using a thick, low density (and usually rather weak) core material, faced on both surfaces with a thin, high-strength skin, or facing.

In order to really be a "structural sandwich", both of these components must be present, along with another very important one, which is a very strong and rigid attachment which holds the facings in intimate contact with the core. The strength and rigidity of this attachment is so important that a structural sandwich made with rubber cement or other soft adhesives does not qualify as a sandwich, even if the core and facings are properly chosen and firmly attached.

Such "rubbery" structures do have a place in the scheme of things, however, and some very good baggage compartment liners are made this way. If you try to calculate their deflection under load, however, the effort always results in a wrong answer, as these structures do not follow the classic formulas for sandwich loads and deflections.

The reason that the sandwich form of composites has been so frequently used lies in the fact that the thin, strong facings can be made to do an outstanding job when they are both separated and supported by the core. To illustrate this, just consider a piece of skin material, such as aluminum sheet metal, and consider its stiffness, strength and weight as having a relative value of one. Figure 3-1 shows how these values increase when a core material is used to make a sandwich out of the same total amount of skin material.

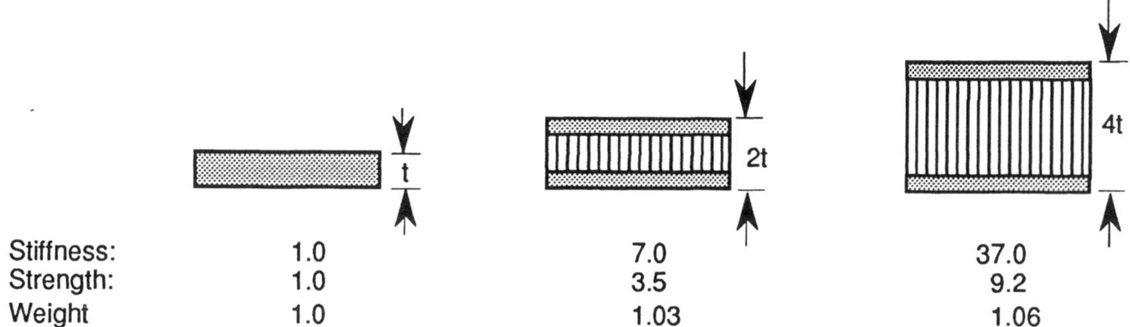

Figure 3-1. Effect of Thicker Core on Stiffness and Strength of a Sandwich Structure

Although the values in the example have been calculated by Hexcel for a highly efficient honeycomb core material and adhesive system, nearly the same results are obtained when one of the low-cost foam or wood materials are used to make the core. A sandwich is simply the ultimate that can be obtained in efficiency for structural members such as plates, beams and large surfaces that have to carry a bending or column load.

Figure 3-1 points out the importance of making the core as thick as practicable, and most sandwiches do have a core that is very much thicker than the facings. Because this method of design uses a large amount of material for core, it also follows that a very light core will save much more weight in a structure when compared to a core which weighs substantially more for a unit of volume.

As a result, it is quite common to compare core materials to each other by quoting the strength at a given density. The strength most often quoted is the compressive strength, as it is the easiest to visualize and to test. The engineers who design with these materials, however, usually also look at the shear strength and the shear modulus. (The term "modulus" is a measure of how much a material will deflect under a given load. A low modulus means that deflections, or bending, will be high, and a high modulus gives smaller deflections.)

In the sandwich structures field, the happy situation is that a great many materials make very good sandwich cores. Let's look at some of them and consider their differences.

WOOD CORES

Some of the most commonly used and lowest cost cores are any of the several light woods. The most popular is Balsa, but substantial use has also been made of Spruce, Mahogany, Redwood, Pine, Fir and many others.

The reason for the popularity of Balsa is its good strength at very low densities. When oriented in the sandwich so that the grain direction is perpendicular to the sandwich facings, its compressive strength is higher than nearly all other cores, even that of all but a few of the high-performance honeycombs.

Balsa cores suffer from having a shear strength that is not nearly as impressive as their compressive strengths, as well as from the usual wood problems of sensitivity to fungus, rot and moisture absorption. (And if you listen to those who don't like wood boats or airplanes, Termites!) Because the material has such wide availability and reasonable cost, it has been used in many commercial, industrial and marine applications.

FOAM CORES

A number of foamed plastic materials are widely used in boat hulls, as well as in home-built aircraft, because they are both reasonable in cost and easy to work with. Each of these material families has a set of mechanical properties, physical properties, working and handling characteristics, and a cost structure all its own. It is nonsense to think one of them will be just like another.

In addition to the differences from one chemical type to another, there may also be major differences between the low-density and high-density members of the same family. A two-pound per cubic foot (pcf) polyurethane will turn out to be vastly different from a 20-pcf piece having exactly the same chemistry and appearance. The resistance to gasoline and solvents, of course, will be about the same since both materials are made from the same polymer.

A few of the more commonly used foam materials are discussed in the following paragraphs, along with some of the characteristics which distinguish them.

POLYSTYRENE FOAM

This is the material used in the VariEze, Long-Eze, Dragonfly, Cozy and most other derivatives of the Eze family, in the wings and control surfaces. It was chosen because its strength properties are well matched to the need of these particular structures, and because it can be hot-wire contoured without generating poisonous gases (like the urethanes do). It has the strong disadvantage of being soluble in Styrene monomer, which is a major part of polyester and vinyl ester resins, and it is softened badly by exposure to gasoline and several other solvents.

The material usually used is Styrofoam, originally produced as a continuous slab by the Dow Chemical Co., but now available from several sources. It is normally a light blue color, a color which is added to the resin to help identify it. This material is quite uniform and reliable in a two pcf density, although the Dow literature also lists a four pcf product.

NOTE! Be careful not to get hold of any "bead foam" material. This is made in many factories by exposing styrene beads, or granules, to live steam while they are contained in a closed mold. The beads expand and bond to each other, but the bond may vary greatly in strength, and as a consequence, the material is completely unsuitable for use as a sandwich core material.

In addition to an unpredictable strength level, the pattern of expanded beads usually leaves an inter-connecting web of tiny spaces which makes the finished material permeable to most liquids and gases after the surface layer, or skin, next to the mold is cut away. Definitely not the material to use in making your Long-Eze wings!

POLYURETHANE FOAM

This is a family of materials which includes both rigid and flexible materials, a great many choices of density and several different chemical versions. Care must be taken in selection of your material, as many of the sources are principally in the cushion, mattress and padding business, and do not understand the requirement of the sandwich core type of application.

Typical problems of the off-breed urethanes are excessive brittleness, variable density within a single piece, internal cracks, soft areas and many others. Fortunately, there are some good suppliers who specialize in the sandwich core supply business and whose material rarely has such problems.

One material in this latter group is Last-A-Foam made by General Plastics. Last-A-Foam is light brown in color, slightly more brittle and possesses better elevated temperature properties than most urethane foams, and is quite tough. Clark foam is still used widely for manufacturing surf boards, but is no longer offered as a precision sliced material for use in sandwich core structures.

Last-A-Foam and nearly all other urethane foams are based on a Polyether formulation. Densities from about 2.5 pcf to 20 pcf or higher are offered, and the supplier makes some excellent literature available to users of the materials.

A couple of words of caution regarding all of the urethane foam materials. First, all except Last-A-Foam are very flammable and all of them, with no exceptions, will produce extremely poisonous gases when burning. For this reason ***YOU SHOULD NOT*** use hot-wire contouring for generating shapes from the material.

The other caution concerns the practice of foaming in place to provide contoured cores. Many boat or plastic supply shops offer two-part, foam-in-place urethane liquids for use in providing flotation in boats. The process seems so simple and obvious that many unwary builders (including some of the largest of the big airframe makers), have tried to make this material work as a structural core for a sandwich having a complex shape.

The success rate has been almost zero. The problem lies in the fact that urethane chemistry is an extremely complex business, depending not just on the two components mixed in the bucket, but also on such things as humidity, barometric pressure, exact temperature of each of the components, as well as the exact temperature of both the bucket and the mold, and many other difficult-to-control factors.

This wide combination of variables makes it rather unlikely that a first-time pour-in-place operation will be successful. It also highlights the fact that, in the composite structures field, much of the success of the final structure often depends upon the exact outcome of a chemical reaction. This step must be carefully and reliable controlled. Where the situation makes this care and reliability difficult or unlikely, the chemical reaction should be handled by your supplier in his factory, with his resident chemist and in-house laboratory right there to make sure the chemical reaction gives the mechanical properties expected in the final product.

One additional caution regarding the urethane materials should be noted. Some suppliers offer very low densities, and the mechanical property curves will often include ranges down to 1 or 2 pcf. The problem is that most urethane foam materials become extremely fragile and somewhat unreliable at densities under 4 pcf, and this figure should normally be considered the minimum for good structural performance.

POLYVINYL CHLORIDE FOAMS

These materials are based on the same chemical family as the familiar garbage bags, plastic pipe and plastic films in common use, but are produced quite differently. The manufacturing process is more difficult than those used to produce most other foams, and the materials are consequently higher priced. Several are available to the amateur builder, most made overseas and imported into the U.S. as finished core materials.

The most commonly used cores of these foams are Klegecell, made in Italy, and Divinycell, whose home is Sweden, with plants in Germany and Texas. A third material, Airex, made in Switzerland, is also available, but is somewhat different from the other two, in that it is quite capable of accepting deformations of up to twenty or thirty percent, without losing its value as a core material. Because of this fact, the material is more suited to heavier boat hull structures than to the lighter structures found in aircraft. Many outstanding boat hulls, some well over one hundred feet in length, have been constructed with this core material in the hull.

Both Klegecell and Divinycell are fully cross-linked PVC foams and make very good structural cores. Their cell size is somewhat larger than most other types of foam cores, and they are not quite so resistant to the wide range of solvents as the urethanes. Densities down to as low as 2.5 pcf are available, and properties at the low end of the range are quite good. Surprisingly, strengths of these foams at densities under 6 pcf are somewhat better than the urethanes, while at densities above 6 pcf the better urethanes usually have higher strength than the PVC materials. This crossover in strength properties is usually not seen in other plastic foam cores.

One occasional problem encountered with the PVC cores, (and sometimes other foams too), has been the infrequent occurrence of an under-cured area in the material. This shows up after layup of a sandwich as a soft area, having been attacked by the laminating resin system. Such instances, while very perplexing at the time, are rather rare.

POLYMETHACRYLIMIDE FOAM

Don't be put off by the impossible name of this one. It is a very high-performance material, known by the trade name of "Rohacell" and available in densities as low as a feather-weight 1.9 pcf. It is extremely fine in cell size and very uniform.

The mechanical properties of Rohacell are well above those of either the urethanes or the PVC foams at all densities, and it is a brilliant white color, much like Clark foam. The materials are resistant to nearly all solvents and chemicals, and are good structures up to well above 250 degrees F. They also have the rather unusual property of expanding during cure of the facings, if the cure temperature is raised to 350 degrees F or higher. This allows a perfect fit between facings and core by simply making an approximate shape for the core and curing the entire sandwich in a closed cavity mold.

This material has no closely equivalent competitor, but is quite generally available from Richmond Aircraft Products, a distributor for Rohm Chemishe, who manufactures it in Germany. Cost of this material is somewhat higher than most other foams, and as a result, it tends to be used in the more weight-sensitive or critical applications. Of all the foam cores, it is the only one which is at all competitive with the Nomex honeycombs in many applications in commercial aircraft.

SYNTACTIC FOAMS

The word "syntactic" simply means that the foam is made by mixing micro balloons, ceramic spheres, or other lightweight aggregate, with a resin system. This results in the density of the material before cure being the same as the density after cure. This is different from most other foams, which are expanded during cure to make the final density much lower than the density of the mixed resin system just prior to the expansion, or "blowing", which takes place as cure proceeds.

These materials may be made by the builder by mixing Q-cells or other micro-spheres with the resin being used, or may be purchased already mixed with a one-part resin system. Many versions are commercially available, but they are so simple to make yourself that most builders just mix them as they are needed. The usable strength of these foams will depend on the resin system employed, the particular type of micro-bubble used, and the ratio of resin to filler in your mix. Since the strongest of the range may be several times as strong as the weakest you might use, it is a good idea to stick with the exact mix specified if you are building to somebody else's plans, or, do some testing if you are striking out on your own.

A WORD OF CAUTION: Some of the bubbles are provided with a surface finish, just as in fibers, and this finish may only be compatible with a single resin system. Be sure the material you are using is intended for use with the resin system with which you are mixing it. Using the wrong resin system will result in very low strength of the final product.

A totally different application of syntactic foam is its use in a thin film, with a very light carrier ply of 106 or 108 glass fabric. for the purpose of providing a smooth surface on a structural laminate part. These "surfacing plies" are now used by many of the aircraft manufacturers to reduce the finishing labor needed to attain a commercially acceptable cosmetic finish on everything from overhead baggage compartments to galley structures. One version of the material has a little blowing agent included in the formulation, so that it can expand to two or three times its original volume, providing a generous capability for filling voids and obscuring fabric pattern show-through.

THE HONEYCOMB CORES

There are about as many members in this family as there are in the foam family of materials, and just as you might expect, the differences between them are rather wide. Most honeycombs are a nested array of hexagonal cells, made from any one of the many standard papers, woven fabrics or metal foils. Many non-hexagonal honeycombs have been used commercially, but the hexagonal cores exhibit both higher strength and a better balance of properties than most other configurations. The details of these core materials are discussed in Chapter 4.

STRENGTH COMPARISONS OF CORE MATERIALS

Comparing the strength of various cores is a little bit dangerous, as the results not only depend on the test method, but also on what has been done to the core during fabrication of the sandwich. If the core is one that is softened by polystyrene, and a polyester or vinyl ester resin system is used in laying up the sandwich, the final product will be much weaker than the values listed in the core producer's literature. Nearly all of the core materials have some such "Achilles Heel" to which they may be sensitive, and the builder must be sure he fully understands all of the many relevant facts concerning the materials he plans to use before attempting to start on his structure.

Assuming this is properly researched, and that these facts are understood and the problem areas avoided, the graphs of Figures 3-2, 3-3 and 3-4 show the relative strength and modulus of some of the core materials mentioned above. These mechanical properties can clearly be seen to increase with increasing density. This is true for all cores, not just for the strength values shown, but for values of all other strengths and for the several modulus values as well. Figure 3-5 illustrates the relative lack of importance of the core modulus in affecting the deflection of long span beams.

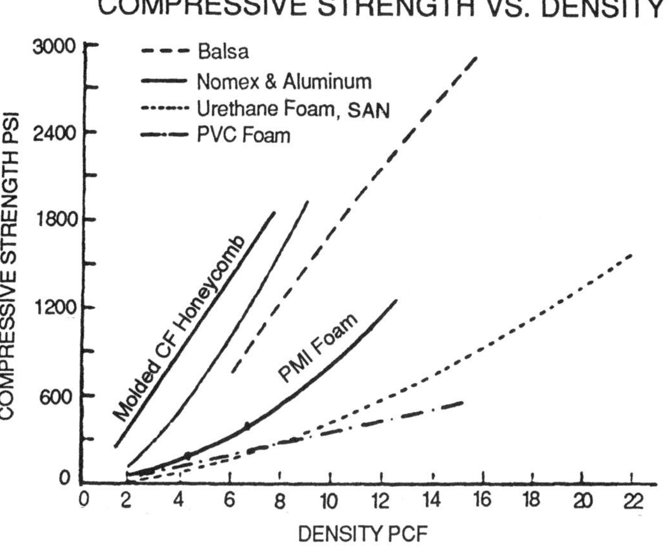

Figure 3-2. Compressive Strength of Several Cores.

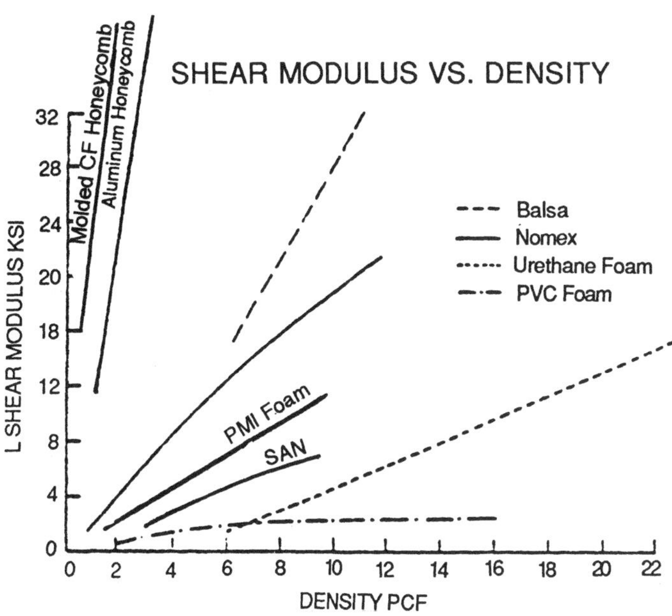

Figure 3.4. Shear Modulus of Several Cores

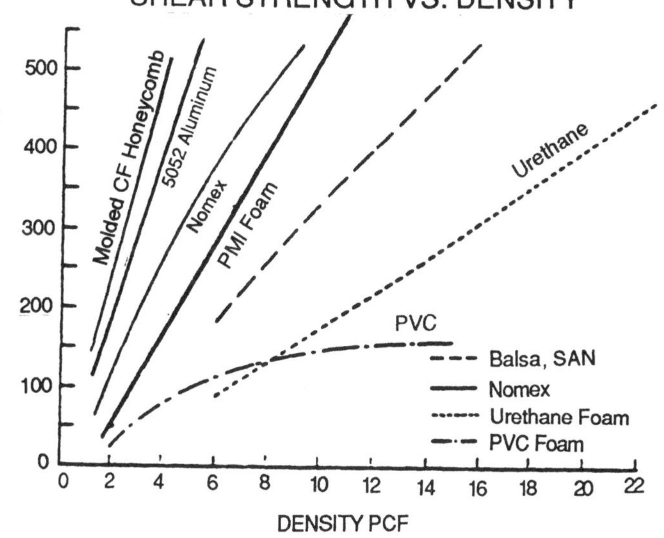

Figure 3-3. Shear Strength of Several Cores

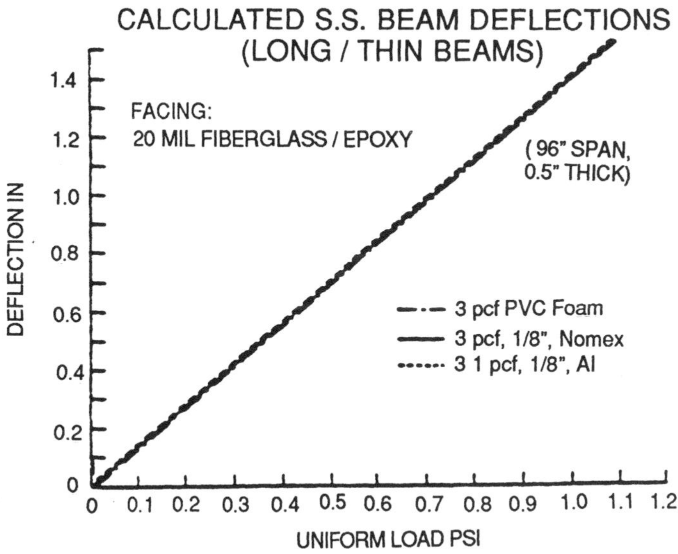

Figure 3-5. Comparative Bending of Beams Using Several Cores.

COST COMPARISONS OF CORE MATERIALS

The cost of the various core materials seems to be as different as their strengths, and placing any reliance on a summary of costs can be a risky business. The real cost is what you actually have to pay for the material when and where you buy it.

Naturally, if you are at Boeing or Airbus and have an annual requirement that approaches $100 million worth of core, the suppliers are going to really sharpen their pencils and offer some attractive pricing. If you are just a private citizen, however, your shipment is not going to justify anybody making a special production run, and you must depend on an Aircraft Spruce or a Wicks to already have the material in stock. Obviously, the price will include both their profit and their cost to buy and inventory the material.

Even with this large discrepancy in pricing, however, the relationship between the cost of different cores is rather stable. The graph in Figure 3-6 attempts to show this relationship. Note that even in the cost chart, only balsa has a cost that is not dependent upon the density of the core. In all other cores, the more material you buy (in pounds), the greater the cost will be.

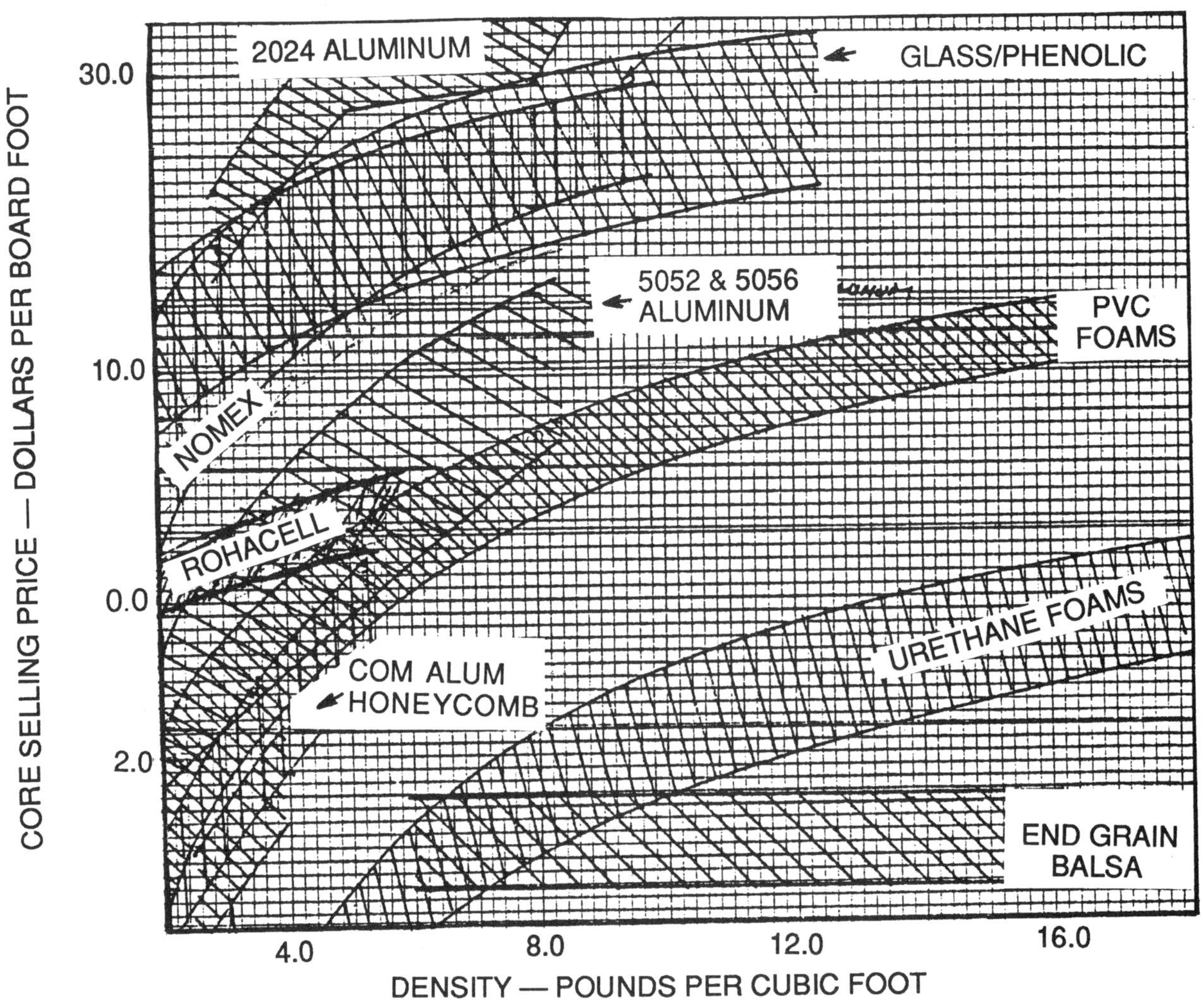

Figure 3-6. Cost Comparison of Cores

CHAPTER 4

HONEYCOMB CORES

There are about as many members in this family as there are in the foam family of materials, and just as you might expect, the differences between them are rather wide. Most honeycombs are a nested array of hexagonal cells, made from any one of the many standard papers, woven fabrics or metal foils. Many non-hexagonal honeycombs have been used commercially, but the hexagonal cores usually exhibit a higher strength and are regarded as having a better balance of properties than most other configurations.

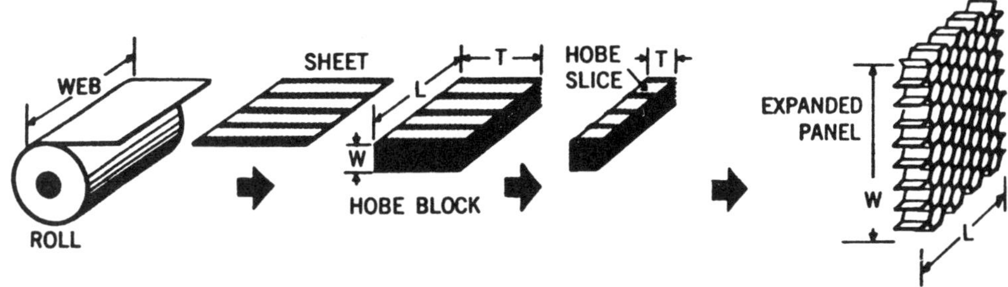

Figure 4-1. Expansion method of producing honeycomb core materials.

The most common methods of manufacture are shown in Figure 4-1. The finished core material is normally furnished in large pieces, usually four feet by eight feet, cut to the exact thickness specified by the user, to a tolerance of a few thousandths of an inch. A few of the various configurations of honeycomb core materials are shown in Figure 4-2 on the next page.

CELL SHAPE

The two primary variables in honeycombs are the material from which it is manufactured and the density of the core. A third important variable is cell configuration. Variations in cell shape arise from problems in manufacture, producing unwanted local or general changes in mechanical properties. Variations in cell shape may also be introduced intentionally, offering some specific advantage over the original configuration.

Some of these variations are so slight as to go unnoticed to the casual observer, but the effect on one or more of the mechanical properties of an errant cell shape can sometimes be all out of proportion to the change in appearance. The only non-hexagonal cores ever produced in substantial production are the corrugated configuration shown in view H of Figure 4-2, and the nested tube cores shown in view F. Since neither of these are often used in structurally critical sandwiches, relatively little specific information is published detailing their structural performance.

Because there are so many additional possible cell shapes as to boggle the mind, we will consider here only the common variations of the hexagonal honeycombs.

VARIATIONS IN CELL SHAPE

The hexagonal honeycombs are produced by either expanding, as shown in Figure 4-1, or by the corrugation process, as shown in Figure 4-3.

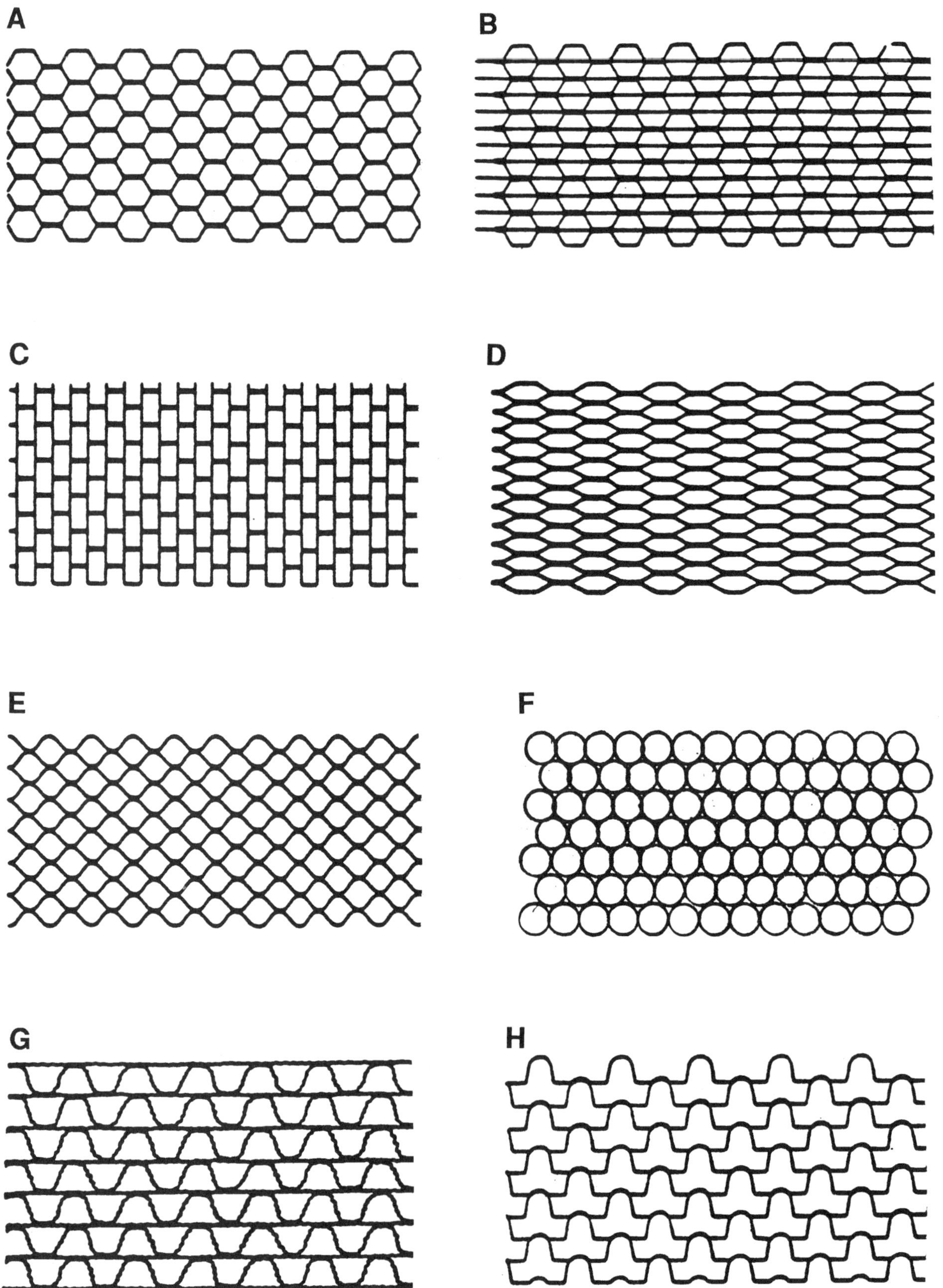

Figure 4-2. Several common cell shapes.

EXPANDED CORE

Since each of these production processes produces its own family of variations and errors in cell shape, they must be considered separately.

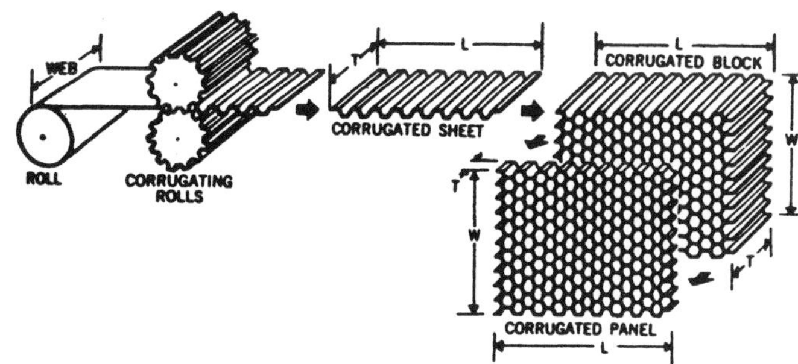

Figure 4-3. Corrugating production method.

The basic cell in both materials can be described as shown in view F of Figure 4-4. A true hexagon would have sides "a", "b", "c" and "d" all of equal length. Such is rarely the case in production core materials. The usual cell shape is designed such that sides "a" and "b" are of unequal length, typically having a ratio of b/a = 0.7. This is done to enable a better balance of strong, or "L" direction, strength and modulus to that of the weak, or "W" direction. This ratio is also subject to some unwanted changes, as the adhesive which holds the ribbons together at the node line, or double-wall portion of the cell, the "b" and "d" sides, is subject to a certain amount of peel-back at the time the core is expanded, as shown in View C of Figure 4-4. In contrast to this situation, a honeycomb made using the corrugation method has no peel-back problem, but instead, has a much larger mass of adhesive contained in the node bond area. This is shown in view D of Figure 4-3. In the Aluminum honeycombs, the amount of peel-back may be nearly zero at a core density of 1 to 2 pcf, but approaching, perhaps, 20% of the intended "b" dimension at a core density of 8 to 10 pcf.

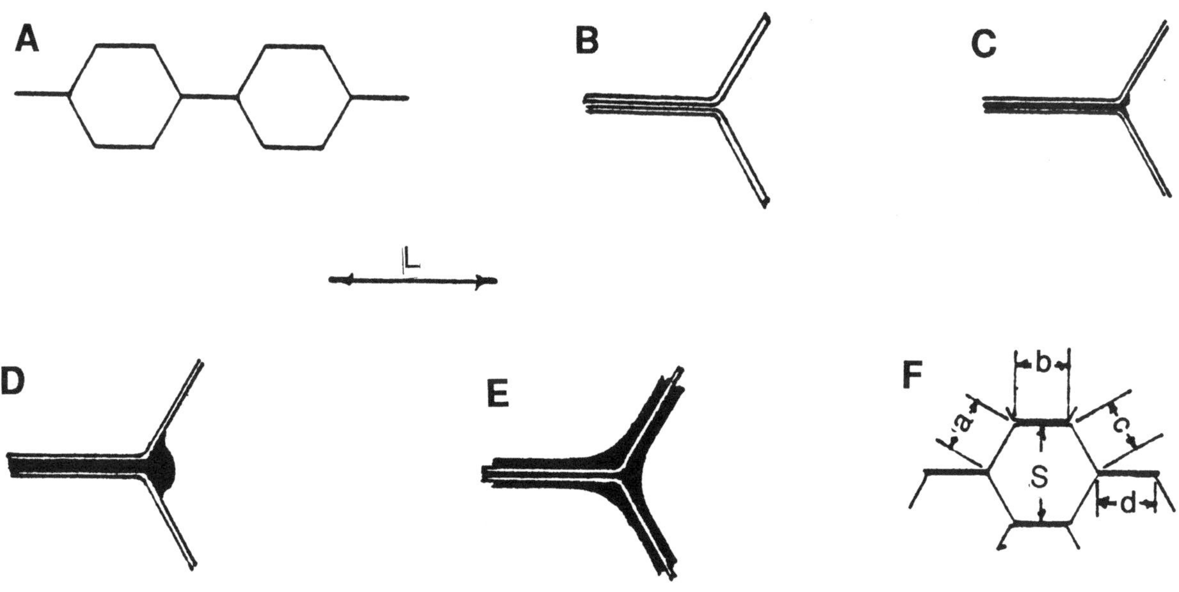

Fig. 4-4. Details of cell construction.

Thus, a different set of print rolls must be used when producing an expanded low density core, compared to the print roll used to produce a high density core. Should the same print pattern be used for both, the low density core will have substantially different characteristics than the high density core made from the same materials. The variability problem is minimized or avoided when a very high peel adhesive is used to produce the node lines, in combination with a special print roll for each density. This minimizes variable peel and assures that the cell geometry will be close to theoretical for all densities.

Many other problems also affect the final cell shape. Those which have little effect on the overall mechanical properties are seen quite often and are usually acceptable. A few such variations are shown in Figure 4-5.

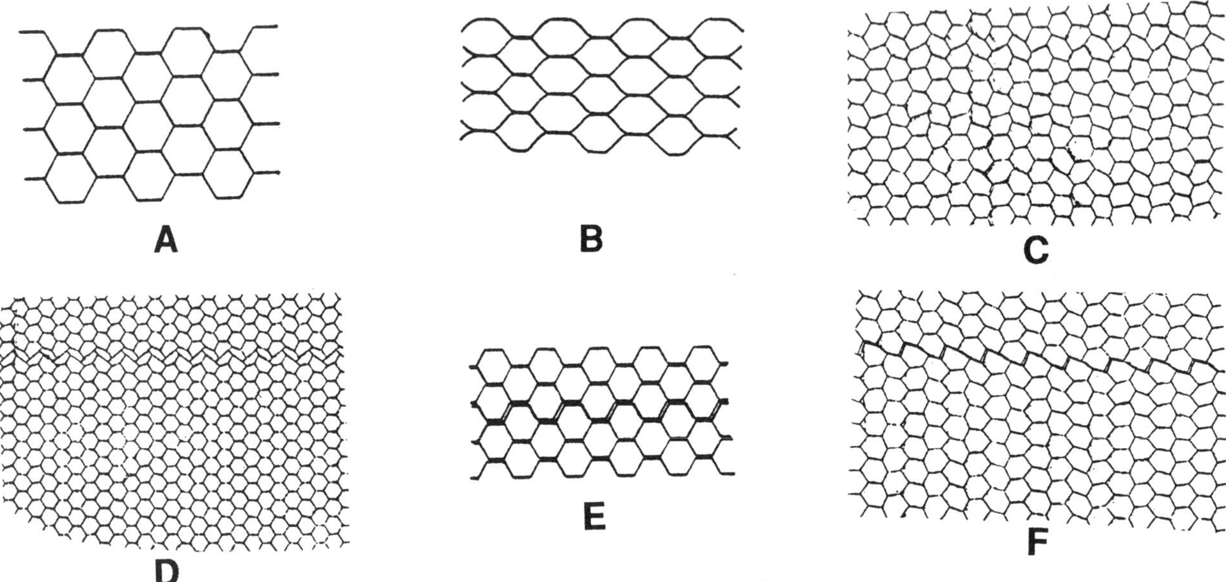

Figure 4-5. Acceptable cell shapes

More severe problems may result in unacceptable configurations. Some examples are the nearly adjacent nested ribbons of Figure 4-6, view A, the offset ribbons of view B, the nested trapeziums, or "teardrop", of views C and D, the brickwork shown in view E, and the herringbone pattern shown in view F. None of these cell shapes is considered acceptable when present to the degree shown.

Since some of these variations cause a serious drop in the mechanical performance of the core at a given density, they are usually not permitted, or only permitted to a limited degree, in the major user specifications. These specifications, however, always allow for a small area of such variations, as the minimum sample size is usually chosen as an area of core at least 12 inches square. Even cell pitch is measured by averaging a 10-cell length of cells.

One of the facts about cell configuration which tends to make the problem less severe, is that the density nearly always increases when the cell configuration is poor. Because the mechanical properties of the core are higher for higher densities, this means that even if the cell shape gives a lower strength, the higher density present in the area of poor configuration makes that low strength figure somewhat higher.

OVER-EXPANDED AND UNDER-EXPANDED CORES

In the case of any normal expanded core product, it is possible to alter the core properties by either under-expanding or over-expanding the core. Since the pitch of the

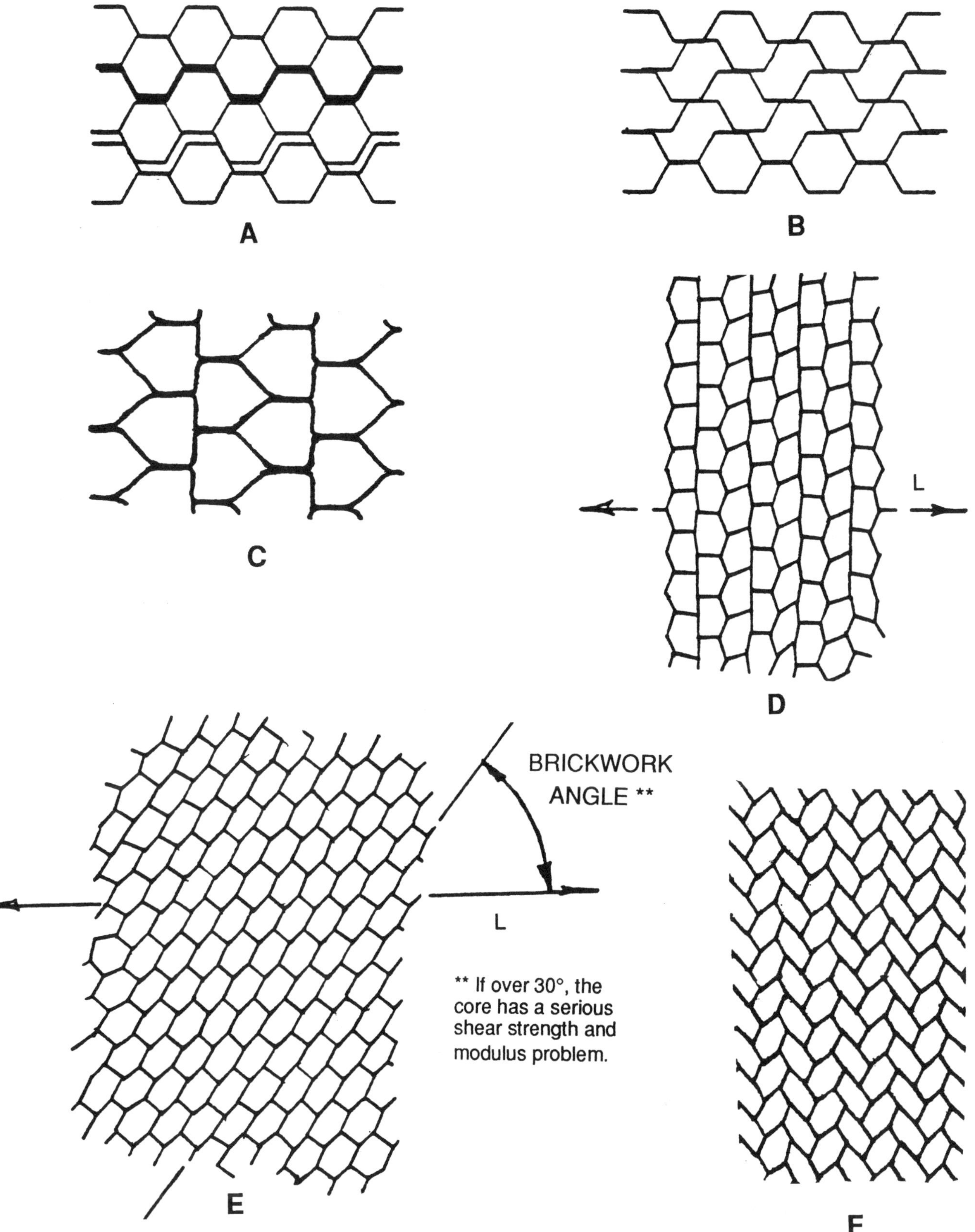

Figure 4-6. Cell shapes which are not generally acceptable.

cells is allowed to vary somewhat in most specifications, a slight amount of this variation is commonly used to adjust the density of the core, if it has been produced at a marginally high or low density. However, substantial changes in the mechanical properties can be accomplished, by large amounts of either over- or under-expansion.

UNDER-EXPANSION

Under-expansion is seldom used in cores other than metal foil based materials, such as aluminum or steel. These materials rely upon the permanent set at the corner of the cell to hold their shape following the expansion operation. If a higher strong direction shear strength or modulus is desired, and if the correspondingly lower weak direction value can be tolerated, it is a simple matter to under-expand the honeycomb to the degree needed to obtain the higher value. This higher value can be easily estimated, since it is roughly proportional to the increase in ribbon count, the number of ribbons per inch of the honeycomb. Thus, if a 25% increase in the shear modulus is needed, the core is under-expanded to produce a 25% increase in the number of ribbons per inch. The corresponding drop in the weak direction strength or modulus is not quite as large as the increase in the strong direction.

NOTE: This expedient should not be used as a substitute for increasing the density of the core at a normal expansion, as the normally expanded core of a higher density will yield an increase in the weak direction properties also. However, it is ordinarily used where a very limited amount of area of increased shear strength or stiffness is needed, as it is available without the need to inventory another core type. View D of Figure 4-2 shows an expanded core having an approximately 40% under-expansion.

OVER-EXPANSION

The change in properties achievable by over-expansion is so remarkable that an entire product line of these materials has evolved over the years. The letters, "OX", are appended to the core designation to indicate the fact that the core has been over-expanded. The usual form of this material is one in which the expansion has been carried out to the physical limit of expansion, so that the "b" and "d" sides of the cell are parallel to each other. The result, as shown in view C of Figure 4-2, is an approximately rectangular shape of the cell, rather than a hexagon.

In addition to altering the relative values of the strong and weak direction mechanical properties, which are reversed, the formability of the slice of raw core is substantially altered. In the place of the usual antielastic curvature displayed by normal honeycomb, the core becomes almost completely formable when wrapped around a cylinder whose axis is parallel to the W direction of the core. The formability of the L direction is only slightly affected by this change in cell geometry, but it, also, becomes somewhat more easily formed. By a happy coincidence, the cost of manufacture of over-expanded core is nearly the same as that of hexagonal core, so the premium charged by the core manufacturer is nominal or non-existent.

CORRUGATED HONEYCOMB

The corrugation process for producing honeycomb, shown in Figure 4-3, results in a material which is superficially the same as that made by the expansion process. The major differences lie in the fact that the corrugated ribbons must each be separately handled and adhesively bonded to each other. The problems of handling and positioning the ribbons accurately increase as the material becomes lighter, such that very low density honeycombs are not very practical to make using this process. Also, the mispositioning of the ribbon in a way which offsets the bonded node, as shown in view B of Fig. 4-6, results in a very easily noted configuration problem, which also results in a serious reduction of shear properties. As a consequence, the degree of such mispositioning, or offset nodes, is nearly always carefully limited to a small proportion of the theoretical node width in the specification requirements of the major users. A maximum of 0.25 times the theoretical node width is common.

The other common problem with corrugated cores is the double ribbon, in which a node offset is so large that the ribbon slips off the node and falls into the open cell below it. This is shown in view A of Fig. 4-6. This is virtually identical to another defect found only in expanded core, which results from placing two identically printed ribbons next to each other, so that the stripes of adhesive fall directly over one another, rather than being offset a half pitch. An example of this is shown in views E and F of Fig. 4-5, with a partial case shown in view D. In either case, this does not result in any loss of properties, but causes an increase in density, along with a corresponding increase in strength. The number of such double ribbons allowed is therefore usually a reasonably large figure, such as one every six or eight inches in the W direction of the core slice.

A third problem encountered in the manufacture of corrugated core, but one which does not usually show up as a defect, is that of holding the total weight of the node adhesive to a reasonably low fraction of the total core weight. While it is common for an expanded core to have an adhesive weight in the node lines of between 0.01 and 0.10 percent of the core weight, corrugated cores ordinarily have a node adhesive fraction of 5.0 to as high as 20.0 percent for the very low density cores. This large adhesive mass is shown in view D of Fig. 4-4.

This large proportion of node adhesive results in the shear properties of the corrugated cores being substantially lower than those of the expanded cores of the same density, with the compressive properties being substantially higher. The end result is that the materials are somewhat different from each other. These differences should be carefully examined before making a choice between corrugated and expanded cores.

MOLDED HONEYCOMB

A recent addition to the several types of honeycomb produced is a novel molded core, made by Ultracor (formerly YLA Specialty Cellular Products). This material appears to be a corrugated core, but is produced by an entirely different process. This process involves laying a sheet of prepreg, consisting of one, two or several layers, over a series of parallel hexagonal bars. The part of the cell where the sheet dips down is filled by another bar along the L direction to complete each ribbon. The entire block of, perhaps several hundred ribbons, is then clamped under pressure in the W direction, and the entire assembly cured at the same time.

Since prepreg layers can be of fabric, unidirectional fibers, randomly oriented mat, or any other array of fibers, a wide variety of end results can be readily achieved. The resulting mechanical properties can be tailored to meet specific goals. Limitations are imposed by the need for a separate mandrel inside each open cell during the cure, but many practical combinations of cell size and materials have been successfully produced and marketed.

In particular, it is possible to achieve much higher shear strengths and moduli with this method than by any other production method known. As a result, the material is used extensively in the manufacture of antenna and solar array structures on various spacecraft. Because of the relatively high cost of production, the process is most practical when using high-cost materials, such as pitch based carbon fiber and some of the very light Kevlar or Zylon fabrics. The biggest single group of these products makes use of a fabric woven from a pitch fiber, which is only 75 grams per square meter in weight. The resulting honeycomb, which comes out at a little over two pounds per cubic foot in density, has nearly twice the strength and modulus of an aluminum or glass based honeycomb of equal density!

DIPPED AND UN-DIPPED HONEYCOMBS

Both expanded and corrugated honeycombs are sometimes dip-coated with resin in an effort to improve mechanical or electrical properties. In the case of aluminum cores, dip coating is rather unusual, as the aluminum alloy from which the foil is produced is inherently stronger than most of the dip resins which might be chosen. Also, the aluminum by itself is reasonably corrosion resistant without any added surface coating. In the case of most non-metal honeycombs, however, a dip coat of some sort is usually provided in nearly all of the standard products. Of all these non metal cores, only Kraft paper is commonly provided without a dip coat. Even this product is dip coated with resin when maximum structural performance is desired.

Honeycomb core materials which are nearly always furnished with at least two separate dip coatings of a structural resin are the structural fabric based materials and the Nomex families of honeycombs. In both of these instances, the properties of the core are so dramatically improved by the dipping process that core manufacturers are quite reluctant to sell the materials without the coating already in place.

The addition of a dip coat of cured resin to a glass fabric or Nomex honeycomb performs several functions. The primary function is to add thickness and stiffness to the thin wall of the cell, sharply improving both the compressive and the shear properties. Improvements can amount to several hundred percent in the case of compressive properties, and upwards of a hundred percent in shear properties, particularly shear modulus. In addition to this remarkable improvement in mechanical properties, an improved resistance to moisture and chemical attack is quite noticeable also.

The dip resin also affords the opportunity to add electrically active material to the honeycomb, which is the basis for the current heightened interest in using honeycomb for structural applications in stealth aircraft. Nearly all of the details regarding this sort of application are classified, and as a consequence, little factual information is available to the public on these materials.

The wall thickness of a typical dipped core may contain considerably more resin as a result of dipping than all of the original material from which the cell wall was made. Also, the corners of the cell tend to pick up more resin than the flat cell sides, which makes the cell appear to be of a rounded geometry, as the higher densities are reached in the dipping process. The amount of resin picked up may be considerably more than the original weight of the core, especially in the case of cores having densities of 9 pcf and higher. A typical cross sectional view of the cell corner of a dipped core is shown in view E of Fig. 4-4.

DIPPING PROBLEMS

Most of the resin in the cell wall has been deposited from dipping and slowly withdrawing from a liquid bath of uncured resin. As a consequence, several problems can arise in the resulting structure. The most common is "agglomeration", a failure of the liquid resin to adequately wet the surface of the cell wall, which causes the resin to draw up into separate droplets of resin, leaving the space between the droplets entirely uncoated. In severe cases, the expected increase in strength is entirely absent, while the added weight is right on target. This defect can be clearly seen, usually with the unaided eye, and is cause for rejection of the honeycomb.

Another problem occurs when the honeycomb block is withdrawn from the liquid resin bath at too high a rate. This results in an uneven coating of resin, in which the side of the core which leaves the dip bath last has a much heavier deposit of resin than the side which was extracted first. In this case, a slice of honeycomb taken from the low-density edge will have much lower properties than a slice of core taken from near the high density edge of the block. This difference may amount to as much as ten or twenty percent, and also can result in a rejectable condition.

A third problem can result from failure to submerge the block with alternate faces of the block each being on the bottom side every other dip cycle. The block almost always has a slight imbalance in the uniformity of the coating as a result of more resin being deposited on the down side, even when the withdrawal rate is quite low. When the block is dipped with the same side down on repeated dip cycles, this difference can reach a total which is high enough to cause a serious difference in density between the top and bottom facing of the block.

To prevent this, the block is usually labeled with an "A" side and a "B" side, so that the dip operator can always alternate the down side with each dip cycle. This problem also mandates that all production dipped honeycombs have an even number of dips, so that the variation in density from one side to the other is minimized. This is particularly important in lower density cores which have very few dips. Even when this sequence is carefully observed, some difference in density can nearly always be detected between slices taken from the middle of the block, as compared to slices taken from a location near the surface of either face of the block.

FLEXIBLE HONEYCOMBS

One of the chronic problems faced by those making curved sandwich structures is that of forcing the core material to conform to a complex shape. Even some simple shapes of constant thickness can present a serious problem when the core is either too rigid, or is fragile enough so that it breaks when forced into a particular shape. The very tough Nomex honeycombs, particularly in the lower densities, can be a great help in solving this sort of problem, as is the switch to an over expanded cell shape for certain part shapes. When a really severe problem is faced, however, one of the solutions often chosen is the adoption of an inherently flexible cell configuration. The core type shown in view H of Figure 4-2 is one of these cell shapes which is in current production and used wherever this problem becomes really serious. The material was made for many years only by Hexcel, but is now also available from Alcore.

KRAFT PAPER HONEYCOMBS

This is the only member of the honeycomb family to have a low enough cost to be an attractive candidate for low cost commercial structures, such as residential interior doors, kitchen cabinet doors, and other such common applications. The problem is that all of these products, even the rather expensive resin-impregnated paper cores, are grossly affected by the presence of either water vapor or liquid moisture. As shown in Figure 4-7, at close to 100% humidity of the surrounding air, even the best of these cores loses about 75% of its dry strength. Since this is considered an additional problem, quite beyond the fungus, mildew and rot problems, the decision is, usually, to avoid use of these materials in any boat or aircraft structures. However, because of the rather attractive cost and outstanding strength/weight ratios, Kraft paper honeycombs are often used in the manufacture of commercial structures designed for indoor use.

GLASS FIBER/PLASTIC HONEYCOMBS

These materials have been in use for over 50 years as cores for radomes and antenna windows, but their high cost has generally restricted broader usage, even in commercial and military airplanes. Many variations exist, and some have remarkable structural properties.

Recently developed materials in this category include both improvements in the shear properties and special, "low observability", cores used in stealth applications. The structural improvements have been obtained by orienting the glass or other fibers at 45° to the cell axis, improving the impregnating resin system used in the base fabric, improving the dip resin, improving the dipping techniques and optimizing the fiber to resin ratio. The use of carbon fiber has permitted development of non-metal cores having approximately the same properties as the best aluminum honeycombs.

High cost usually rules out the use of these materials for any but aerospace applications where their special attributes make them worth the price.

NOMEX ARAMID PAPER HONEYCOMBS

This family of honeycomb materials is based on the Nomex paper made by the Dupont Company. It is a very unusual material due to the fact that it makes a remarkably tough and damage-resistant sandwich, even when the core is at less than 2 pcf, and the facings are a single ply of glass cloth 0.010 inches or less in thickness.

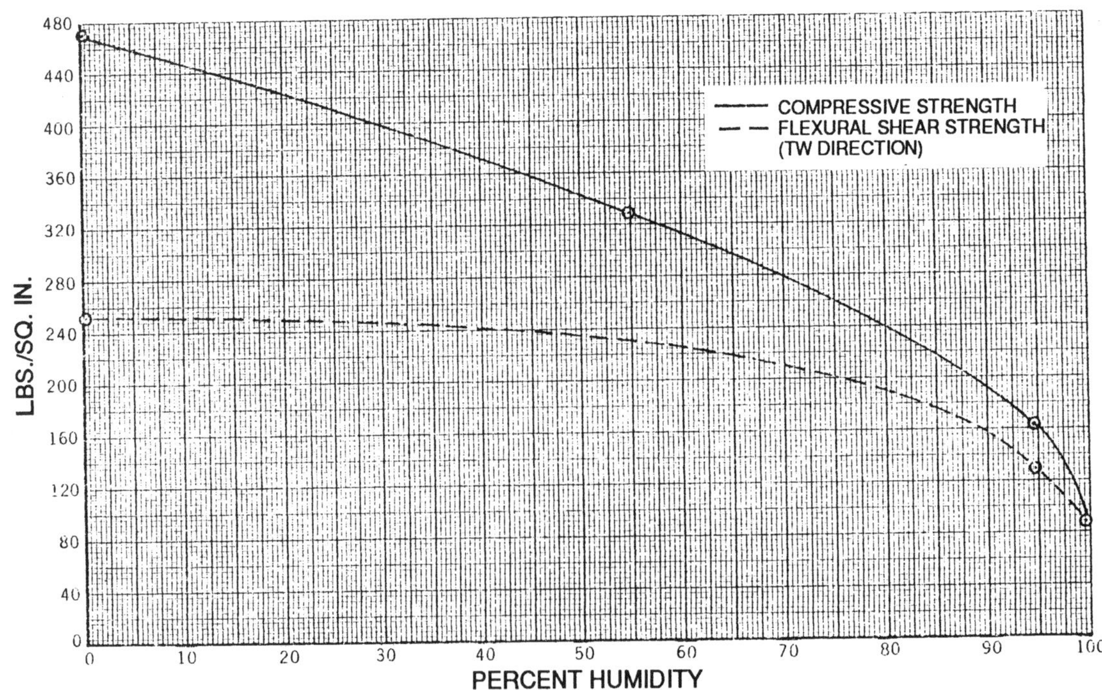

Figure 4-7. Effect of humidity on Kraft paper honeycomb core.

That particular combination makes up a large part of the cabin interior structure of nearly all large commercial aircraft. Even in the floor panels the core density is only 4 to 6 pcf, except in the aisles of the larger planes, where densities up to 9 pcf are sometimes used to resist the wheel loads imposed by food service carts. The extra effort required in achieving a good skin bond is probably the only thing that will continue to stand in the way of more substantial usage by small aircraft builders and boat builders, although the high strength-weight ratios available should make them much more common in the next few years.

KOREX ARAMID HONEYCOMB

This is a product quite similar to Nomex honeycomb, but which employs much higher strength and higher modulus Kevlar fibers in the manufacture of the paper web. It is also a product of the Dupont company. It was first introduced in 1992, and is being currently promoted and recommended for applications which require higher strength, lower moisture pickup and/or lighter weight than can be provided by Nomex cores. The difference lies almost entirely in the higher strength, modulus and lower moisture expansion than is normal for either Nomex or Glass fiber based honeycombs. It is even superior to the bias oriented glass fiber cores. Except for these substantial improvements, the core looks and handles very much the same as the widely used Nomex honeycombs.

Korex honeycomb is higher in cost than Nomex honeycomb, and is expected to see service only in spacecraft and high performance aircraft applications, in much the same way as Kevlar honeycomb.

CARBON FIBER AND KEVLAR HONEYCOMBS

Such materials are available, but their very high cost severely limits their use. At this time, they are commonly seen only in space vehicle structures and a few specific applications in the larger and newer transport aircraft, where the higher cost is considered a fair trade-off for the high specific strength or electrical performance (of Kevlar) which they offer.

One new application for the pitch based carbon fibers is their use as the core material for sandwich panels which require heat to be transmitted from one facing to the opposite facing. This was initially used only on the Boeing Model 777 interior nacelle panels as mentioned in Chapter 1, page 1-9. Recently the decision was made to use a similar construction for the new Model 737-700 and -800. This honeycomb makes use of the high thermal conductivity commonly provided by pitch fibers in the range of 75 msi to 120 msi modulus. The volume of core required by these three models made the entire honeycomb industry snap to attention. However after a few years a cost study at Boeing resulted in a return to the original choice, aluminum honeycomb.

STAINLESS STEEL, TITANIUM AND SUPER-ALLOY HONEYCOMBS

Both welded core materials and all-welded sandwich are available in these materials. Cost, however, is very high when compared to the more common honeycombs (upwards of $1,000 per square foot for many combinations!), and usage is limited to jet engine parts and some hot areas of military aircraft.

CHAPTER 5

CARVING AND FORMING CORE MATERIALS

The steps in going from a rectangular slab of a core material to a finished shape suitable for the making of an aileron, a fuselage shell or a boat hull are usually considered a separate subject. The whole thing can be quite confusing at first glance, but if the problems are taken one by one, you will find that each is rather simple.

CUTTING PLAN DIMENSIONS FROM SLICES OF ANY CORE

Nearly all core materials cut well using either a long-blade bread knife with a serrated edge, or a band saw. If the plan form is complicated it is best to overlay the core with self-adhesive paper, such as ordinary shelf paper. This way you can lay out the lines to be cut while keeping the surface of the core clean. If you have a lot of straight cuts to make, it helps to set up your work table so that it has a straight slot just wide enough for the knife blade. The slot should be exactly 90 degrees to the edge of the table or 90 degrees to a guide bar, to enable a perfectly cut square corner.

For thicker slices of core, you will find that the cutting remains easy, but that it becomes quite difficult to hold the cut edge perfectly square. For this situation, you can resort to either a table saw or band saw, or make another special fixture to hold the knife blade vertical as you make the cut. The only thing that will defeat you on this one is if the core density gets too high and the knife blade starts to deflect while cutting.

As long as the core density is under 5 pounds, and the thickness is not over an inch or two, these methods usually work very well. At thicknesses of 4 to 5 inches, however, most any core is difficult to cut by hand, especially if you expect perfectly square edges.

It is important to achieve a good, true cut on the first try. Subsequent attempts tend to make the piece look worse, and each handling contaminates the surface of the core more. In the case of most honeycombs it is extremely difficult to remove a tiny bit from an edge and leave a true line, either in planform or in perpendicularity. All the foams and end-grain balsa cores are much easier to handle in this regard.

MULTIPLE CUTS OF THE SAME SHAPE

When many pieces need to be cut to exactly the same shape, such as a triangular gusset of core, or a rib section, it is simple to make a steel rule die, just like the big guys.

Steel rule stock is usually available from any printers' supply house, or from any large print shop. It is about .032 inches thick, 3/4 inches wide, and sharpened continuously along one edge. It is furnished in a continuous length, many feet long, coiled up in a flat package, just like band-saw blade stock.

To make the die, lay out the desired shape on a piece of 3/8 inch or 1/2 inch thick plywood and cut along the line with a coping saw, or a thin-blade band saw. The steel rule is formed by bending it so that it fits exactly into the cut line, with the sharpened edge sticking out toward you. This assembly is backed up with a solid piece of 3/4 inch or 1 inch thick plywood, and the whole thing is set over a slice of core and put in an arbor press. The sharp edge of the steel rule will thus cut right through the slice of core, yielding a cleanly cut piece the exact shape as that laid out on the first piece of plywood. Naturally, a piece of plywood, Masonite or plastic is needed for the sharp edge of the steel rule to bottom out against when it cuts through the core.

A small, simply shaped die made like this can be produced in 20 minutes, and using a hand arbor press, you can make about 20 parts per minute. If you plan more than just a few parts, however, you should also include a piece of sponge rubber or a spring in the middle of the shape, so that the cut piece of core will be pushed out of the die when you raise the press ram. This method is usable for large parts also, but the press can get expensive. One low-cost alternative I have used is to put the press bed surfacing piece of plywood on the floor, and then use the impact of two people jumping on the die to provide the cutting force. The parts made in this manner were the "orange peel" sections used to make the nose radome on the F-86D fighter aircraft. They were about 36 inches long, and at that time the company did not have a press big enough to hold the tool.

SCARFED OR BEVELED EDGES

Where a constant thickness slice of core must have a sloping edge, it is wise to set up a guide to hold the angle constant, whether using a saw or knife. Holding such an angle by eye is usually not satisfactory, unless you are a **remarkably** good workman. If the beveled edge is to be cut along a planform that is not a straight line, it is wise to use adhesive paper on the surface of the core, so that the line to be followed can be seen clearly. Remember that a poor outline on the beveled edge of the core will show as a poorly finished part, even though the strength and serviceability may be quite acceptable.

Where a half-round, or other specially shaped edge detail is desired, a common router usually can be used, provided the router bit will produce the shape to be cut. A honeycomb core can be a serious problem, as most honeycombs must be filled with a solid material, such as polyethylene glycol, or ice, if the edge is to be cut by this method. The woods and foams can be routed easily, however, without any preparation except for a template to guide the router along the edge.

CONSTANT THICKNESS CURVED SHAPES

Virtually all of the foam core materials can be precurved to nearly any shape imaginable by using heat-forming methods. This does not work on wood cores nearly as well, unless steam is added to the process. Several steps in the process deserve comment:

1. The temperature needed will be different for each core material, and sometimes for different densities or different production lots of the same core material. These temperatures will be as low as 200 degrees F for urethane foams, up to perhaps 325 to 400 degrees F for the Rohacell foams (polymethacrylimides). Some experimentation should always be done, even if you already have experience with a foam, as slight differences in the formulation will give noticeable differences in the best forming temperature for a given foam.

2. Allow plenty of time for the foam to get up to temperature. Too short a time will leave the center of the slice at too low a temperature to allow forming without breakage. Even for thin slices, a half hour or so is none too long. (An exception is the case where the part is small enough to fit into a microwave oven. The heat-up cycle can then be a minute or less, but must be determined by experimenting.)
3. Provide some tooling or a method to force the desired shape into the slice of foam rapidly, as it will be cooling all the while you are trying to form it. The exact amount of time you have to complete the forming will depend on the thickness of the slice (thinner ones give you less time), and the core density, (heavier cores give you more time), as well as the cold wind blowing across your workplace. Usually, however, you will have a minute or so to do the forming, and it is not much of a problem.

The really major exception to the statement made above is the time limit required for Nomex or glass fiber/plastic honeycomb cores, which have a low density and a very high heat transfer capability. In this case the usual maximum time that can be allowed between removal from the oven and completion of the forming operation is about **two seconds!** Needless to say, there are not very many firms performing this operation on a routine basis, other than the core manufacturers themselves.

Where the forming is rather complex, or where exact conformance to the desired shape is important, it is usually best to use a female mold and a vacuum bag, putting the whole assembly into an oven to allow the slice to be creep formed. This takes much longer, but lower temperatures may be used, and accurate duplication of complex shapes is easily accomplished. Using Clark urethane foam as an example, 24 hours at 200 degrees F will produce a nearly perfect shape. Obviously, both the bag material and the female mold must be capable of withstanding the temperatures and pressures involved. Nearly any vacuum source may be used, as only a couple of psi are needed to make the method work well.

If you use your household vacuum cleaner, however, be **sure** that it is not a type that uses the vacuum tank exhaust to cool the motor. If it is, it will burn it out quickly. One of the best vacuum sources can be bought at most any hardware store. It is a little gadget that attaches to a garden hose, and uses a venturi to provide suction for draining water out of cellars and basements.

CARVING STRAIGHT-LINE CONTOURS

Where the shape to be carved can be defined by straight-line elements, the core cutting can be done by using either a hot wire, a band saw, a router traveling along a straight track, or sandpaper glued to the edge of a straight two-by-four.

HOT-WIRE CARVING

This is one of the most commonly used methods for carving foam. It was originally set up by Burt Rutan for the Vari-Eze surfaces, as well as those of his subsequent designs, for parts of Quickie's and Dragonfly's wing surfaces and for the many follow-on designs from other designers. Simply glue or pin an accurate template to each end of a block of foam and follow the template carefully, one element line at a time, with a tightly drawn, resistance-heated wire. (Simple in theory, but a little more difficult in practice.) If you have provided a good, light-weight frame to hold the wire, a good, sensitive power control device to provide the exact heat on the wire desired, and a Nichrome wire instead of stainless, so that it doesn't stretch so much when hot -- and if you and your partner both have steady hands, steady nerves and true eye sight -- the results will be beautiful.

Perhaps the most frequent source of problems is not the carving itself, but the accuracy, or lack of it, in the base table, the original block of foam, the templates, and in the exact positioning of the templates on the core block ends. The underlying fact is that whatever shape you carve will be the shape that the final part takes, including the actual surface finish. As in many other facets of composite construction, it pays to take great pains and do a careful job at every step, starting with leveling and truing the table, so that the subsequent steps will come out as planned.

A WORD OF CAUTION:

The hot-wire method works on almost any low-density core material, especially foams, but some of these materials give off some rather deadly fumes when the hot wire burns the foam cell walls.

Also, some of the foams are easily set afire and are therefore hazardous around a red-hot wire. The end result is that only styrofoam is considered suitable for this carving method. Any other foams should be avoided, if you plan to use a hot wire.

Anyone planning to use the hot-wire carving method would be well advised to get a copy of the video cassette once put out by the Rutan Aircraft Factory, or, more recently, by the builder of the record-holding AR-5, Mike Arnold. The video you want is his #2, "How It Was Built," a very detailed account of the exact methods, including a substantial amount of detail on the hot-wire carving process. The Rutan video covers the details of the methods used in building the Long-EZ.

SANDPAPER BEAM CARVING

This sounds awful, but works well, as long as the material is rather low in density. Nearly all the foam cores up to densities of about 6 pounds can be easily shaped using this approach. This method requires gluing a hard, abrasion-resistant rib, usually a precured piece of fiberglass, to each end of the foam block or assembly to be carved. This abrasion-resistant rib serves as the template, defining each element line of the shape to be carved, and, if designed carefully, may also be left in place to form part of the final structure.

The foam is usually hacked to near the final shape with a butcher knife or saw, so that not too much material remains to be removed by sanding. The sandpaper is glued to the edge of a very straight two-by-four, and sanding is done by drawing the sanding surface back and forth across the end templates. As with the hot wire method, the accurate prelocation of the end templates, and working only along single element lines from one end to the other is important. Also, as in the hot wire method, it is important to achieve a good quality surface finish on the carved surface, because of the influence this surface has on the final part finish.

In either of these methods, hot-wire or straight beam sanding, it is important to be sure you are setting **both** ends of the line on the **same** element line point. For example if you have the element lines numbered, do not set one end of the beam or hot wire on number 23, while the other end is on number 26 or some other number. If you do, you will carve a saddle into the part, where the middle of the foam is much lower than the ends.

If you are sanding a hollow box, rather than a solid piece of foam, as in the old Polywagon wing, you should remember that this box will deflect away from the cutting surface of the sandpaper if you proceed too aggressively with the shaping. It is therefore a good idea to practice the art with finesse, rather than force. The same is true of shaping a curved piece of foam that is temporarily attached to a framework, as in the Dragonfly fuselage. Any over-aggressiveness pushes the work around to a point that, when you are done, it will sigh with relief and move part way back, leaving a surface shape you didn't intend, and which is not at all the shape you thought would be there.

BAND SAW AND ROUTER CARVING

Most carving done in the field of commercial and military aircraft manufacture is performed by one of these methods. For example, all of the honeycomb cores for the main rotor blades for all of the Bell Helicopters produced from about 1956 to the present day, have been router-carved. The method, however, is not easily adapted for use in a home project, as the equipment and tooling are rather expensive. The actual cost to produce the carved parts is quite low, once the initial setup is completed.

The router bits used in this method are unique in that they employ a slitting saw, similar to those used on a milling machine, but held in a chuck that looks like a valve stem from an internal combustion engine. As a consequence, they are universally called, "valve-stem cutters". The shank diameter is made to the correct diameter and style to use the router chosen. Nearly any router will work, as long as the shaft is true, the bearings tight, and the speed at or above 20,000 rpm. A drawing of one version of this tool is shown in Figures 5-1, 5-2 and 5-3.

In the case of band saw carving, the process is limited both by the straight-line-element requirement and by the maximum length of an unsupported band saw blade. Surprisingly, saws with an unsupported blade length of 10 feet or more have been common in the industry. In all cases, the blade tension, wheel balance, blade guide condition, blade weld quality, and many other factors make these band saws much different from the average shop saw. One surprising note, however, is that blade life is usually about as long as that of any other band saw, and sometimes even longer, since the material being cut is often a very low density honeycomb.

FULLY SCULPTURED SURFACES

These surfaces are those that cannot be formed or described by a series of straight lines or a body of rotation. They are simply sculptured. Typical are wing tips, wing root fairings, and NACA inlet ducts, or any duct entrance that cannot be turned on a lathe. Surprisingly, these shapes are those in which the person with a shop in his garage need take a back seat to no one.

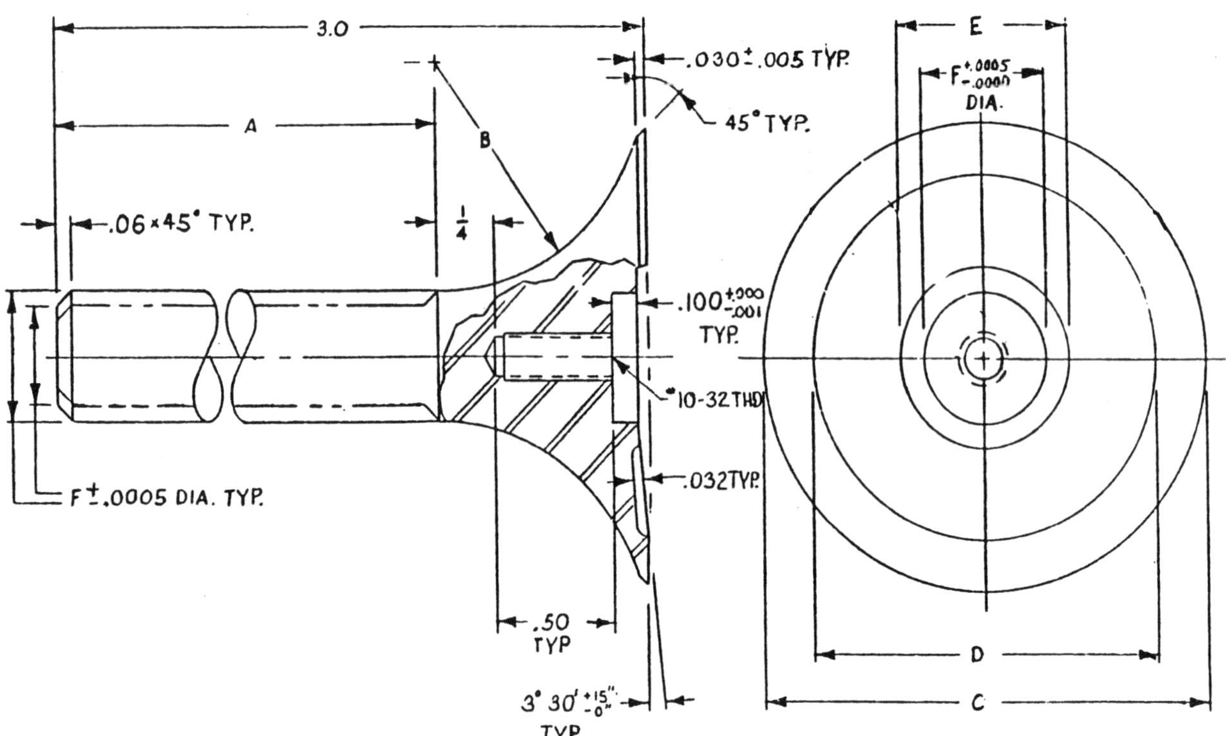

Figure 5-1. General arrangement of a valve-stem cutter used for carving of all honeycomb materials. The drawing shows a family of cutters of different sizes. Figures 5-2 and 5-3 show additional details. (Courtesy Bishop J. Moll, Honeycomb Consultant.)

The usual production approach is to describe the shape mathematically, program a tape for a five-axis mill, and then set up a production run. The true shape actually desired, however, is often slightly different from the mathematically simple description, and a master model is carved by hand. A tape is then produced by hand tracing the model with a duplicating tracer hooked up to a computer. The production carving machine then duplicates this hand-carved model, but not exactly. Errors in surface location usually run from .005 inches to .020 inches or more, and can make for many subsequent production problems.

The homebuilder, on the other hand, just makes his hand-carved part out of the core he intends to use, finishing until he is satisfied with the contour. He then applies the facings, again sanding off any unwanted thickness build-up in order to reach the perfect final part shape. It is likely that the part thus produced by a careful builder is not merely the equal of a factory-produced part from a major airframe company, but may actually be superior in conformance to the shape sought.

This is one of the factors that amazed the NASA group, who found that Burt Rutan's wing surfaces were truly laminar-flow most of the way to the trailing edge. (I have watched Burt's people working on these surfaces, and I can assure you that he does only what he tells his builders to do. It's just hard to believe that they are that meticulous and methodical in achieving the exact shape they are looking for.)

The methods the homebuilder must use to achieve these sculptured shapes are just

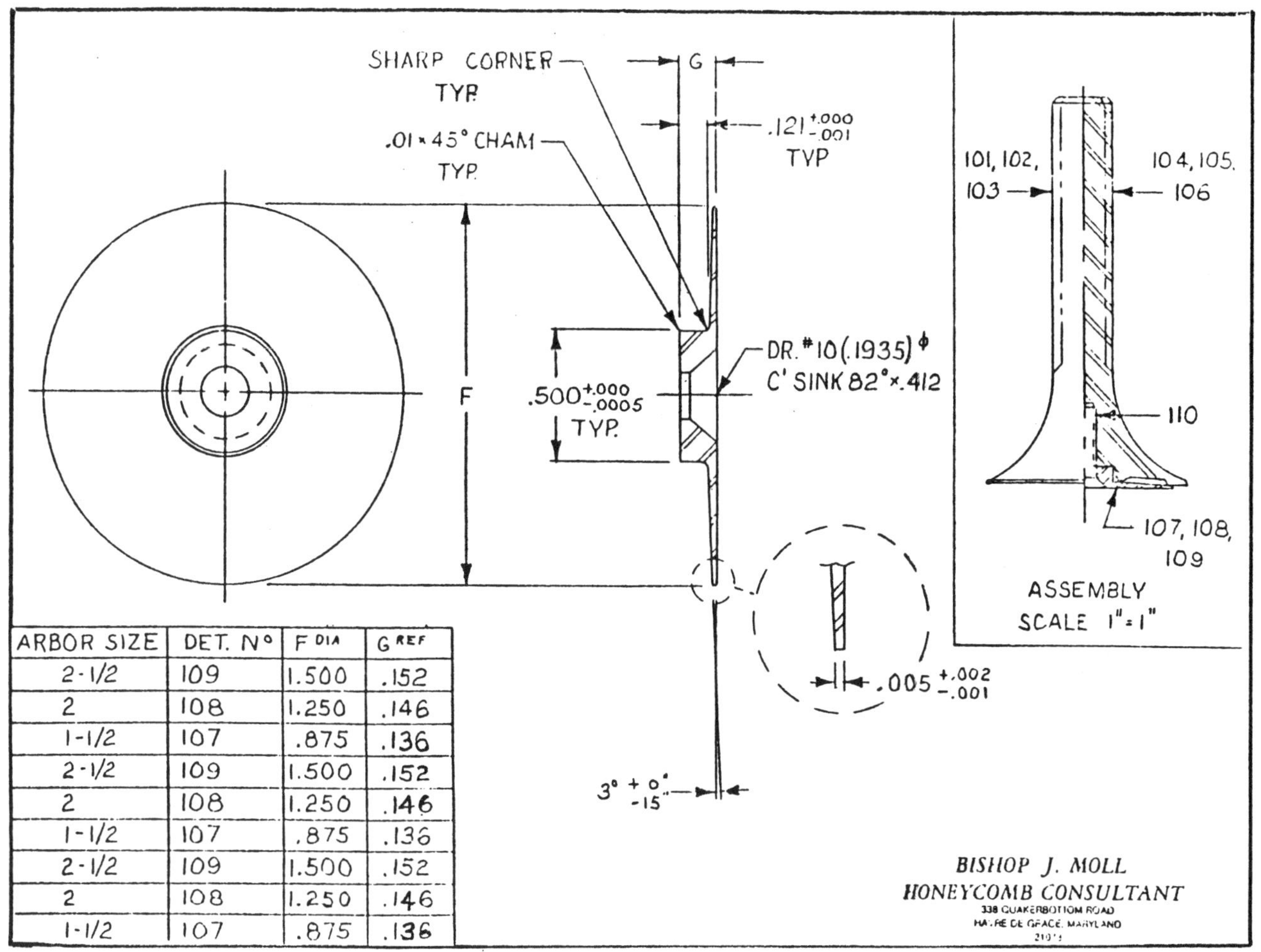

ARBOR SIZE	DET. Nº	F DIA	G REF
2-1/2	109	1.500	.152
2	108	1.250	.146
1-1/2	107	.875	.136
2-1/2	109	1.500	.152
2	108	1.250	.146
1-1/2	107	.875	.136
2-1/2	109	1.500	.152
2	108	1.250	.146
1-1/2	107	.875	.136

Figure 5-2

DET	REQ'D.	DESCRIPTION	MATERIAL
110	1	FLT HD CAP SCR #10-32 × 5/8 LG. STD	
104-109	1 EA.	BAR	4130 COND. F
103	1	2-1/2" SAW ARBOR ASSY. DET. 106. 109, 110	
102	1	2" SAW ARBOR ASSY. DET. 105, 108, 110	
101	1	1-1/2" SAW ARBOR ASSY. DET. 104, 107, 110	

ARBOR SIZE	DET. Nº.	A REF	B RAD	C DIA.	D DIA.	E DIA.	F DIA.
2-1/2	106	2.095	.875	1.750	1.375	.687	.500
2	105	2.345	.625	1.500	1.125	.687	.500
1-1/2	104	2.595	.375	1.125	.875	.687	.500
2-1/2	106	2.095	.875	1.750	1.375	.687	.375
2	105	2.345	.625	1.500	1.125	.687	.375
1-1/2	104	2.595	.375	1.125	.875	.687	.375
2-1/2	106	2.095	.875	1.750	1.375	.687	.250
2	105	2.345	.625	1.500	1.125	.687	.250
1-1/2	104	2.595	.375	1.125	.875	.687	.250

Figure 5-3

what the name implies. You must use hot-wires, knives, saws, Surforms, rasps, chisels and lots and lots of sandpaper. The result will reflect both your skill as a sculptor and your patience and perseverance as a workman. Fortunately, you can make up for any lack of expertise as a sculptor with additional patience and attention as a workman.

Just don't assume that the shape you want will be quick to arrive at. Take time and work slowly, and you will be amazed at the quality of the work you can turn out. In addition to patience, of course, those who are successful at it seem to spend a lot of time making very accurate templates at a great many stations.

HOT-FORMING FOAM

An unusual method of producing small sculptured shapes is to hot-press a foam core. The Rohacell materials will work very well, and some of the Divinycell materials as well.

In this method, a split female mold is prepared, having the exact shape that is required of the final piece of core. A somewhat larger piece of core is then heated above its softening point (determined by trial and error) and forced into the desired shape by closing the halves of the mold on it. The assembly is then left to cool down to room temperature, and the core removed.

In the case of Rohacell, the foam can also be used as a source of pressure for curing the sandwich assembly, even without the pre-compression step. What happens is that the core begins to expand when it reaches a temperature somewhat above 350 degrees F, and when used with a 350 curing resin system, results in a tightly fitting core assembly which has provided the cure pressure for the facings. Because the core will readily expand at these temperatures, the exact starting shape of the core is of very little consequence. This method is often used in small trailing edge shapes or other small, closed shapes which represent a very difficult tooling problem for conventional carving.

MULTIPLE PIECE CORES

When a piece of core is quite large, and must be contoured in a single piece, it is common to use several different pieces of core glued together to make the final piece. Particularly when making full-depth wing sections by hot-wiring styrofoam, we see more than a single piece of core in a single finished structure. When this method of building up the final size of the core is used, do not fail to glue the pieces together!

Core is normally a very lightly loaded structural part of the sandwich structure, and it is easy to forget that the loads the core carries are very important. If two adjacent pieces of core are not structurally attached, the juncture is not capable of transmitting shear loads through the core, and they must be carried by the skin in secondary bending. This means of carrying the load is very dangerous, as it generates peeling forces which tend to separate the skin from the core, thereby causing a general failure of the structure in the area. Regardless of whether you are using a thin slice of core in a boat hull, or a thick piece in the wing of your "Cozy", **DO NOT FAIL TO BOND THE PIECES OF CORE TOGETHER! FAILURE TO DO SO WILL INVITE STRUCTURAL FAILURE.**

CHAPTER 6

THE RESIN MATRIX

Composite structures can consist of literally any strong fiber, carried in any matrix from hard candy to Titanium. Most of the commonly used combinations consist of fibers encapsulated in a resin system based on polymers of polyester, vinyl ester, epoxy or polycyanate. The purpose of the resin system is to transfer load from fiber to fiber, and sometimes, to carry the load from fiber composite to an insert, edge member or fitting. The choice of resin will depend upon a great many factors.

Some of these may be:

- the structural requirements of the part,
- the cost pressure on the project,
- the mixing, handling, lay up, tooling and curing facilities available
- the type of service environment expected,
- the aerodynamic or cosmetic surface requirements and,
- the length of service life expected or desired.

Nearly all of the resins used in composites are a combination of several resins and various additives. Hence, the term **resin system** is often used instead of **resin**. Such additives can raise or lower the viscosity, change the resistance to ultraviolet radiation, improve the interlaminar shear strength as well as the toughness, increase the strength of resin-rich areas in the laminate, increase or reduce translucence of, or add color to the laminate, and change the surface tension and/or the wettability during the low-viscosity period just before cure.

As many as 20 or more of these modifying materials may be added to the basic resin **before** catalysts, promoters or hardeners are added. The obvious consequence is that the resin system user must be both sophisticated and capable in resin chemistry if he expects to successfully further modify this resin system. *DON'T DO IT!*

Another term often heard as a substitute for resin is **matrix resin**, usually used when the formulation is intended for use in a composite laminate. The term, "resin system", is much broader, and may include systems formulated for use in one of many other applications, such as:

- for gluing composite pieces to each other after one or both have been cured (as an adhesive),
- for gluing metal parts to composite parts (as an adhesive),
- for potting or encapsulating small and delicate electronic assemblies (as a potting resin),
- for forming an impervious and attractive surface film (as a paint),
- for forming an impervious interior surface on a gasoline tank (as a sealant),
- for providing a solid cured area in a core material, into which a hole may be drilled and tapped for inserting a threaded fastener or through bolt (as a core filler),
- as well as countless other specific and important uses.

The exact formulation of the resin system usually varies from one application to another, in order to obtain the best performance for use in that specific application. Thus, when the system is optimized as an adhesive, it may not be very good when used as a matrix resin or potting resin. Therefore, the manufacturers and specialists who develop and report the final structural properties of the material as used will tend to refer to the material as a **resin system** when it is intended for some particular job.

POLYESTERS vs. VINYL ESTERS vs. EPOXIES

When comparing the three most commonly used resin materials, one of the first things that becomes obvious is that most of the composite structures in current production employ a matrix resin system based on a polyester polymer. Such products as motor and sailing boats up to 100 feet or more in length, showers and bathtubs, industrial ducting, blower housings and storage tanks all are heavy users of fiberglass/polyester materials. The two compelling reasons for this are low cost and simple, reliable chemistry.

Both orthopthalic (the lowest cost type) and isopthalic polyester resin materials are produced in huge quantities by several large suppliers. They sell, in quantity, for around $5 per gallon. This is low when compared with a typical price of vinyl esters at $12 or a formulated epoxy at about $35 to $60.

The curing mechanism for both of these polyester resins, as well as for the more expensive vinyl esters, is the addition of chemicals that cause the resin molecules to lengthen and solidify. The reaction is so straightforward that small errors in the amount of hardener added will only affect the time the system needs to develop full strength. This tolerance for error on the part of the user makes for a reliable and highly practical cure system.

Neither the ortho- nor the iso- version of the polyesters is suitable for use where higher strengths or low final weight is a primary goal. They do not develop the high shear strength levels available from many of the vinyl ester or epoxy systems. The vinyl esters, however, share the simple curing mechanism of the polyesters and also are relatively low in viscosity, making them easy to use. The vinyl esters also bond better to the Kevlar and carbon fiber reinforcements than do the polyesters, and they have higher strength as a cured, neat resin (which means no fibers, fillers or microballoons in it). The cost of these resins, although higher than the polyesters, is well below even the lower cost epoxies.

Several vinyl esters are commonly used as matrix resins. Typical are Derakane 411-45, 470-36 and 510-40, all made by Dow Chemical Company, Midland, Michigan, and Atlac 580-05, made by ICI Americas Inc., Wilmington, Delaware. These resins all contain 36% to 50% Styrene, which contributes to their low viscosity, but also makes them somewhat toxic, the same as the polyesters. In addition, the resin is usually mixed with a promoter, Cobalt napthenate (CONAP), an accelerator, dimethyl aniline (DMA), and a catalyst, which may be methyl ethyl ketone peroxide (MEKP), benzoyl peroxide (BPO) or cumene hydroperoxide (CHP).

These several materials are not only complicated to use, but can also be VERY hazardous. In addition to being poisonous, the MEKP will explode violently if mixed directly with the CONAP or DMA. The usual procedure is to mix in the promoters ahead of time and store the resin already promoted. Then when a part is to be made, the MEKP is added to the promoted resin. The chemical reaction that causes the liquid resin to solidify and strengthen then progresses rapidly.

Figure 6-1, which is reprinted from the Derakane literature, "Fabricating Tips", provided by Dow Chemical, shows the effect of differing amounts of these chemicals on the time the resin takes to "gel" or solidify. The time to gel can be easily adjusted from about 20 minutes to an hour. In addition to this flexibility, the new Dow literature describes the use of 2-4 Pentanedione (acetyl acetonate) in extending pot life to as long as 8 to 10 hours! At the end of the extended period, the resin gels quickly and develops even higher strength than unextended samples!

AMBIENT TEMPERATURE	**COOL (55-70° F)**	**WARM (65-85° F)**	**HOT (80-100° F)**
% CoNap (6%)	0.4	0.3	0.2
% Dimethylaniline (100%)	0.15	0.1	0.00
% MEKP (60%)	1.0-2.0	1.0-2.0	0.75-1.25
5 GALLON ACCELERATED			
Derakane Resin	5 Gal	5 Gal	5 Gal
CoNap	67 cc	50 cc	33 cc
Dimethylaniline	24 cc	16 cc	0 cc

Note: Replacement of depleted dimethylaniline may be necessary to maintain gel time. Shelf life is several weeks.

When withdrawing samples for catalysis, they should have a uniform light purple tint.

CATALYSIS			
Master Batch	1 quart	1 quart	1 quart
MEKP	9-18 cc	9-18 cc	7-13 cc

Figure 6-1. Derakane 411-45 and 510-A-40 Mix Variations for Different Shop Temperatures.

Figure 6-2 shows the effect of a change in shop working temperature on gel time of Dow's Derakane 411-45. From this information, it can be seen that you can also adjust the speed of the curing reaction to suit the actual temperature in the work room, so that the working time available comfortably fits the time needed to complete the layup.

In addition to the data on the variation of gel time with different amounts of catalyst, promoter and accelerator, this manual also contains some specific data on the use of a thixotropic agent for thickening the resins, so that they do not run so much in the layup. Many other details on the use of the resins is also included. All of this information should be used with caution, however, as it assumes that the reader is a competent formulator.

The term, "gel-time", is commonly used for most resin systems to describe the time which elapses between when the resin is mixed with catalyst and when the mixed resin

(Minutes)	10-20	20-40	40-60
TEMPERATURE, °F			
Cool — 60's	2.0% MEKP .3% CoNap .2% DMA †	1.5% MEKP .3% CoNap .1% DMA	1.0% MEKP .3% CoNap .05% DMA
Warm — 70's	1.5% MEKP .3% CoNap .2% DMA	1.0% MEKP .3% CoNap .05% DMA 1	.75% MEKP .2% CoNap .05% DMA 1
Hot — 80's	1.25% MEKP .2% CoNap .1% DMA	1.0% MEKP .2% CoNap .05% DMA 1	.75% MEKP .2% CoNap

† DMA = Dimethylaniline
1 May be left out at the higher temperatures

Figure 6-2. Derakane 411-45 and 510-A Resins. Gel Times for Various Mixes and Shop Temperatures.

batch begins to thicken, or "gel". Since this point is about the same as that point at which the resin is becoming too thick to easily saturate the dry fibers in the layup, it is almost universally used as the measure of the pot-life of the mix. Actually, the mix can be used slightly longer in most cases, but these minor differences are usually ignored, and the gel-time is quoted as the pot-life. This generality is used on all of the resin systems, whether they be polyester, vinyl ester, epoxy or even one of the various others.

Although the amount of material to be added to polyester or vinyl-ester resins is shown to an accuracy of one-tenth of a percent, the actual measurement at the time of mixing is easily done using a calibrated hypodermic, a medicine dose cup, or even the cap from the container in which the material is furnished. The proportioning thus turns out to be far easier in practice than it sounds from a written description.

Another advantage of both polyesters and vinyl esters over the epoxy systems results from the cure mechanism. The former families harden by polymerizing, which is the lengthening of the ester molecules. The epoxy systems harden by cross-linking two different materials (the A and B parts of the mix). Hence, small errors in the amount of promoter or catalyst added to the ester systems will cause the materials to gel too fast or too slowly. In either case, the ester materials can normally be counted on to harden to a full cure without problem, as long as there is time to lay up the part, and ample cure time is allowed.

EPOXY RESINS

In the case of the epoxy resin system, however, the cross-linking method of hardening means that there must be the correct amount of B molecules so that all A molecules have something to which they can cross-link. In most systems, relatively small errors in the proportioning or subsequent mixing of the two parts of the system will result in a serious shortfall in the final strength of the part. With the epoxies, this strength never achieves the expected value if such an error has been made. A slow-curing vinyl ester, however, takes longer to cure, but nevertheless does eventually reach the expected strength level.

With all their problems, why use the epoxies at all? The reason is that they usually offer more strength, stiffness and toughness than any of the other systems in common use. There are several hundred epoxy resins commercially available, and several times as many hardeners. One manufacturer, Pacific Anchor, gives a wall chart to formulators using their hardeners. The chart lists 64 manufacturers of just these hardeners, making a total of 72 different hardeners for epoxies. Not all of the manufacturers make all of the types listed, but many of the squares on that chart have a trade name listed. This gives us a total of 500 or more hardeners to combine with a couple of hundred resins, for a total of perhaps one hundred thousand epoxy/hardener combinations.

Obviously, no one can test and report on all of these combinations, but a surprising number are in daily use for some application or another. For our purposes, we will consider only a few of the epoxy systems, as most are not well suited for use in the manufacture of high performance structures in a home workshop.

The liquid system, **"Safe-T-Poxy,"** is now superseded by Epolite 2427, which is made by H. B. Fuller. Although the name, "Safe-T-Poxy", is no longer used for any product, the name seems to survive and is recognized by most of the supply houses, who will furnish the newer material when asked for the obsolete one. This system, along with the "West System", materials, produced by Gougeon Brothers, are two of the best of the many room-temperature curing, low viscosity epoxies suitable for hand lay-up of dry fabric.

The mix ratio of these materials is forgiving enough so that small mixing errors can be accommodated, but as is true of many of these systems, they exhibit poor creep strength at elevated temperature. Elevated temperature, in this case, means the temperature reached in the upper facing of foam-cored structures when exposed to the sun with the wind not blowing. Skin temperatures produced under these conditions may reach 190 to 210° F when the surface is painted a dark color, and nearly all the room-temperature cure systems, as well as the 250° curing systems, with a few exceptions, show a substantial loss in strength until the temperature drops.

You should plan on taking two steps to help the situation. One is to paint the structure white, which dramatically lowers surface temperature. Figure 6-3, on the next page, shows surface temperatures reached by foam cored structures on a summer day in Michigan. Temperature rise for various colors was compared in these tests. The other step which can be taken is to post-cure the structure after the room temperature cure has been completed.

Nearly all composites made with a room-temperature cure system will benefit, either modestly or greatly, from this step. Both the creep strength and normally measured static strength values can be expected to improve at least 5% to 10%, with some systems showing far greater increases. This post-cure may be done by painting the part black and then setting it out in the sun, or even better, by building a temporary oven of plywood or cardboard and then heating the part to 160 to 200 degrees F for ten or twenty hours. Just be sure you don't burn your shop down! Refer to the resin manufacturers recommendations for post cure information on a specific resin system.

A second group of epoxies includes the many prepregs. A prepreg is a fabric or a layer of unidirectional fibers, pre-impregnated with a resin system which has all of the catalysts, promoters and additives included, but which uses a curing system which allows an extended time at room temperature in which a part can be laid up, bagged and made ready for cure at a higher temperature.

The various prepregs are offered with cure temperatures as low as 150 degrees F, and several are suitable for vacuum-bag molding with only an oven and a vacuum pump as shop equipment. Although these systems are not in common use in home workshops, due to the high cost of building an oven and maintaining freezer storage space, they offer excellent mechanical properties and nearly complete relief from the serious health problems associated with liquid resin systems. The resin in the prepreg has been B-staged, or partially cured, and almost none of the resin material is transferred to the hands. In spite of this, protective gloves are still recommended when working with the materials, since some of them contain very toxic additives, and any of them may become contaminated from the normal oil present in your hands. An added benefit is

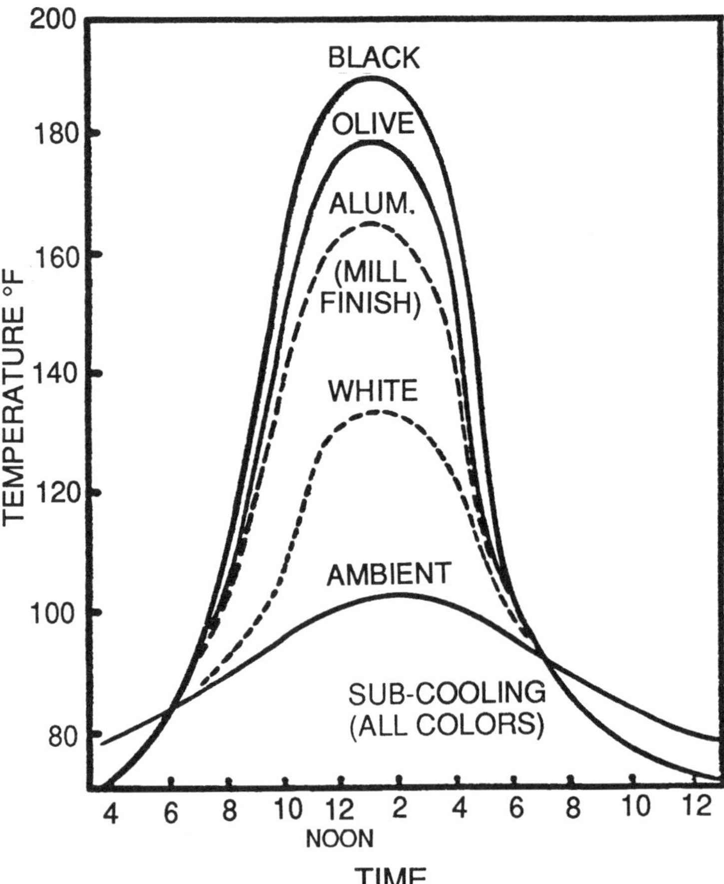

Figure 6-3. Daytime panel surface temperatures for aluminum face, foam core sandwich structures. (Courtesy Dow Chemical Co.)

that prepregs may be ordered with any degree of softness, drapability and tack, or, "stickiness", desired.

A third group of epoxies available to boat, racing car or aircraft builders is that typified by Hysol EA 934 (actually an adhesive), or Ren RP4002 through RP4033. These materials are characterized by their ability to be cured at room temperature, and then, after a suitable post-cure, exhibit high structural strength at service temperatures as high as 450 degrees F! Figure 6-4 shows the performance claimed by the manufacturer for Hysol EA 934.

SOME STANDARD CAUTIONS

One problem associated with nearly all uncured resins, including both vinyl esters and epoxies, liquid or prepreg, is that they tend to absorb either moisture or carbon dioxide from the surrounding air. These absorbed materials then inhibit the polymerization or cross-linking reaction, and the finished composite never achieves the expected strength levels. Therefore, it is usually recommended that parts not be laid up at temperatures lower than 60 degrees F. If layup at such a low temperature must be done, it is recommended that the laid up but uncured parts be moved into a warm room — or an oven — immediately after layup for faster cure, thus exposing the resin to the contaminating atmosphere for much less time. When an extended pot life additive is used, shop humidity should stay below 40% to 50%, as the extend-life resin still has the same susceptibility to absorbing inhibiting gases.

In any case, if the cure proceeds into the second day without substantial progress, you may never get a part, as the inhibition of the cure is irreversible. To help this situation, you can take two steps.

EA 934 — TYPICAL CURED PROPERTIES

Tensile Lap Shear Strength per ASTM D-1002, after curing 7 days at 77°F. Chromic acid etched 2024-T3 Alclad, 0.063 inch thick, 0.05 inch overlap, one inch wide:

Test Temp. °F	-432	-67	77	180	200	250	300	400	500
Strength, psi	3,180	2,600	3,100	2,200	1,800	1,500	1,000	760	400

Trend of Tensile Lap Shear Strength vs. exposure at 300°F & 400°F after curing one hour at 200°F.

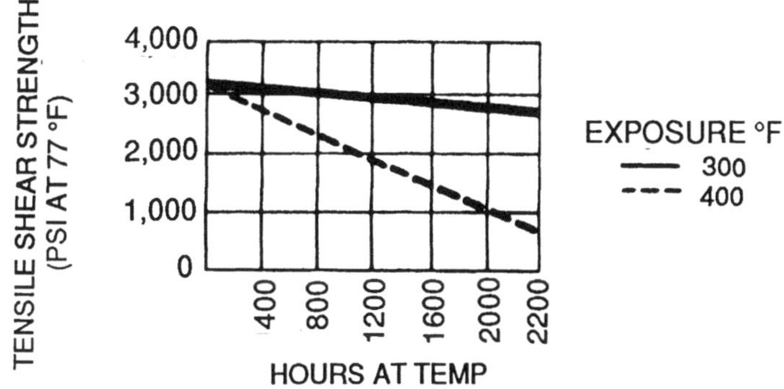

Figure 6-4. Typical cured properties of Hysol EA934 Resin.

- The first step is to have both a thermometer and a humidity indicator in your work area. Do not mix resin or lay up parts if the temperature drops below 60 degrees F, or if the humidity rises above 50%.
- The second is to lay a plastic film over the as-yet uncured part, to minimize the absorption of any inhibitors from the surrounding air.

The degree of cure which the system finally reaches can be easily checked with a Barcol or Shore hardness tester, using the resin manufacturer's minimum hardness data as a lower limit to indicate a complete cure.

Another problem which may occur with nearly any room-temperature cured system is the exothermic reaction, or **"exotherm"**, a rapid increase in temperature of the resin as the cure gets under way. With some systems, the energy that is given off by the cure reaction can actually boil the still-liquid resin and ruin the part. Very large parts may even set themselves on fire! The problem is worse for larger quantities and usually puts an upper limit on the thickness of the part that can be laid up at one time, as well as the largest batch of resin which can be mixed for use at one time.

Shrinkage is also a problem in most resins, but varies considerable from one type to another. The polyesters shrink as much as 5% or 6% in every direction during cure, while the vinyl esters and epoxies shrink somewhat less. Because of this high shrinkage, polyester resins may be used for permanent tooling only if allowance for this shrinkage is made. Even after the parts are completely cured, polyester tools will continue to shrink and change shape as time passes. For truly permanent tools, a good grade of tooling epoxy must be employed, or the tools will continue changing shape forever.

One additional standard caution on the use of epoxy resins concerns the fact that many excellent epoxy adhesives and laminating resins may not bond well to wood surfaces. This problem can lead to serious consequences if ignored. One of the solutions when bonding to wood is to use one of the West System materials, as they were originally developed for laminating glass fabric to wooden boat hulls, and their use completely avoids this compatibility problem.

ONE FINAL CAUTION, and the most important of all: All uncured resins, catalysts, promoters, hardeners, curing agents, and all solvents should be considered poisons and should be handled as such.

The actual degree of danger differs for each, but all are toxic to some significant degree. Most give off harmful vapors, or solvents that can be readily absorbed through the skin. Epoxies have a cumulative effect, whereby after several years with no problem, you suddenly become sensitized and, even at a distance, develop a painful rash or other serious effect.

To avoid these problems, you should set up your shop so that a fan moves fresh air across the work area and **OUT** the door. Do not allow it to be recirculated, even in winter. The increased heating costs are well worth the added benefit to your long-range health.

Use one of the many barrier creams available to protect your hands, so that any contact with resin will be much less significant. Use rubber gloves as well as thin cotton gloves under them. Sweaty hands can aggravate skin that may already be breaking out in a rash as a result of exposure to your resins and solvents.

DO NOT EVER wash your hands in solvent after working with the resins. The solvent absorbs the skin's natural oils and drives the poisonous resin materials right through the skin and into your system. Instead, use one of the waterless hand cleaners, then wash your hands thoroughly, but **only with soap and water.**

Safe-T-Poxy, mentioned because it is the most common name for home use epoxies, originally had a relatively non-toxic rating, which led to the adoption of that unfortunate name. Nevertheless, it can and does cause skin rash and various other problems. Therefore, all of the usual precautions should be observed, even when using what you believe to be the original Safe-T-Poxy. As a matter of interest, Hexcel actually discontinued the use of the name, "Safe-T-Poxy" a number of years ago, in an effort to reduce the chance of a user misunderstanding the danger of exposure.

One step that is often neglected, even in commercial shops, is to **READ the Material Safety Data Sheets** supplied with all adhesive and resin materials. If your supplier has neglected to include them, request them at once. Every manufacturer is required by law to provide them for each resin material he sells, and is obligated by law to furnish them, without additional cost, to users of his materials.

If you are interested in digging deeper into the safety considerations of using composite materials, a good reference work has recently been published by the American Composite Materials Association, commonly known as ACMA. The title is, *Safe Handling of Advanced Composite Materials Components: Health Information.* It is 30 pages long, and is available at a cost of $15 from ACMA, 1010 North Glebe Road, Suite 450, Arlington, VA 22201. Their telephone number is (703) 525-0571.

One last thing to consider, just as with core materials, is where to go for the best information. Manufacturers, even those that only sell their materials through distributor organizations, usually will supply technical information direct to the user, free of charge.

Dow Chemical Company, for example, will gladly send you a packet of information on their Derakane vinyl ester resins, as well as all of the Material Safety Data Sheets. The same is true of 3M, Hysol, Hexcel, H.B. Fuller, the Ren Plastics Division of Ciba, Gougen Brothers, and all the other resin suppliers.

NEW RESINS

Like some of the very advanced fibers, new resin systems are under constant development, with new candidates being announced with increasing frequency. Most of the developments are addressed at providing higher service temperatures for the finished composite or adhesive bonded structures, largely in response to needs of the armed services. However, one new class (actually a very old class) of resins is that of the "Thermoplastics". These include many resins used in other more simple applications for decades, such as nylon, polyethylene and polystyrene, as well as newer ones, such as polyolefins and polypropylene.

In this new approach, these resins are introduced to the dry fibers by adding them in the form of commingled plastic threads, or dry powder, so that a quantity of fiber or fabric will carry with it the proper amount of resin to make an optimum structure after the resin is heated to a liquid state and then cooled to form the matrix structure.

The great attraction of this system is that it completely avoids the need to handle any liquids, pastes, or catalyst, as well as offering the possibility of greatly shortened time in the molding process. A finished part can be produced in only the time required to heat and cool the assembly, with no time at all required to allow completion of a chemical reaction. This leads to the possibility of hot-stamping formed parts from flat sheets of finished composite, much as sheet metal parts are now stamped out in a hydraulic press or drop hammer. In addition, most of the resins considered are lower in cost than the traditional liquid or paste systems, and have no toxicity problems. Much work remains to be done, however, before these materials and processes become commonplace.

CHAPTER 7

MOLDING METHODS AND TOOLING

Let's look at the steps that are necessary to make a part or a complete structure of the composite materials discussed in the previous chapters. This is a procedure referred to by composite manufacturers as "Process Engineering."

The first thing to consider is the resin mixing operation. The two general categories of materials call for two separate approaches. For the vinyl esters, most of the proportioning of the liquids may be done by volume measurements, using a small syringe or a calibrated hypodermic needle to measure out the catalyst to be added. The barrel of a hypodermic needle can be marked off to as little as 0.1 cc. Larger amounts can be measured using the small graduated glass tubes with markings down to about 1 cc, or containers that are only marked with total capacity, whether that is 10 cc, one pint or two gallons.

Some of the reagents used in the hardening systems can be purchased with a calibrated bottle cap indicating the correct amount to add to a quart, a pint, or other common size of resin mix. Some reagents come in tiny bottles with a dropper-tip, in which the drop of material squeezed from the bottle has been calibrated to be the correct amount for some other small amount of resin, such as a half-pint. The added convenience of not having to weigh or accurately measure each part of every resin mix justifies looking into and adopting such minor conveniences, because these steps will be repeated many times during the project's construction phase. Be sure that you get the complete resin mixing instructions for the system you are using from the resin manufacturer. They are normally more complete and comprehensive than the directions you will get from most kit suppliers or material outlets.

A typical case is Derakane 411-45, used in the original Glasair family of aircraft, as well as many others. Dow Chemical, who makes these resins, will send you an information package on this material that has many different combinations of resin, promoter and catalysts, covering various pot lives at various shop temperatures, as shown in Figures 6-1 and 6-2 in the previous Chapter. This information is very useful if your working conditions vary somewhat from those that prevailed when the kit builder wrote his instructions. Also, the resin manufacturer inevitably knows more about the product than anybody else, and should be considered the primary source for such information and advice, rather than a distributor or kit builder.

In the case of an epoxy resin system, the volume-measurements commonly used in the ester-based materials are usually not accurate enough to ensure both structural reliability and uniform working life for the mix. It is important that the builder obtain or make for himself a reliable and accurate balance for correctly proportioning the resin and the hardener. An attractive alternative is the purchase of a proportioning dispenser pump, available from several of the material distributors.

Again careful attention should be paid to this piece of equipment, as it will affect both the convenience and efficiency of your work, as well as the strength of your finished pieces of structure.

A basic piece of equipment that should always be on hand is a beam balance, which is accurate to at least 0.1 gram or less. It is usually available and reasonably priced at a large drug store. To augment this, you will want a proportioning balance that is set for the exact ratio of the resin and hardener used in your system. If you can obtain a set of plans for the Rutan VariEze or Long Eze, you will find a good one included. These will work quite well if you do a good job of constructing the balance, and check it from time to time against an accurate scale.

When you are using the epoxy systems, you **must not** vary the proportions to adjust the viscosity of the mix, even when it seems to be like molasses. Some diluents, such as glycidyl ethers, dibutyl phthalate, or even toluene are used by the formulators to make a lower viscosity mix, but these should be avoided by the homebuilder, as they will substantially change the mechanical properties of your final part. In addition to this structural problem, some diluents are very toxic, and may cause far more serious problems than the one you are trying to solve. Homebuilders who are serious enough to attempt this sort of thing should read the appropriate sections of *Handbook of Adhesives,* by Irving Skeist. It covers many of the technical aspects of the materials and offers several specific formulations.

MIXING RESINS

One precaution has overriding importance. **Mix the resin carefully and thoroughly!** A poorly or partially mixed cup of resin will result in parts which develop far lower than expected strength levels. This also applies to the material left in the cup. Pour the mix back into the first cup after mixing in the second cup, so that all of both the A-component and the B-component will be used in the mix. When stirring the resin, try to trap as little air in the mix as possible. Be careful of drill-motor mixers, as some of them produce an aerated mix containing thousands of tiny bubbles that remain suspended in the mix. These bubbles will cause the cured resin to fall far short of the expected mechanical properties. (Air has much lower strength than cured resin!) The only corrective action you can take is to either avoid creating them as you mix, or to subject the mixed resin to a vacuum to extract the air after it is mixed in.

This problem is not so tricky in the vinyl esters, as both the resin and various materials added to the resin are rather low in viscosity. As a result, trapped air comes out easily and the proportion is not so critical. Pay attention to the mixing problem, as any defect developing at this point automatically becomes a permanent part of your structure.

EXOTHERMIC REACTION

A side effect of the cure is the exothermic nature of the chemical reaction, the generation of heat, usually just called "exotherm". Every combination of chemicals used as structural matrix resin systems produces an exotherm—some more, some less. If your mixing instructions don't offer comments on what to expect, it is a good idea to mix a batch in the largest mix volume you will use at one time, and measure both the increase in temperature and the time available before the viscosity starts to rapidly increase. If the exotherm is high, you must decrease the size of the mix in order to maintain control. Also, it is best to perform this check out-of-doors, as the temperature rise of some systems is so rapid and goes up so high, that the mix can catch fire and burn briskly.

PREPARING THE FIBERS

The above comments regarding mixing resins assume that fibers are ready to use. However, much work is usually required to ready the fiber in order to apply the resin. In most cases, you will be using fibers in the form of a woven fabric that comes in a roll. If the project is a large one, you will probably have more than one weave style of fabric and more than one width. Most fabrics are available in widths of 38 inch, 50 inch, 60 inch, and 72 inch, and in some of the more popular weave styles, down to 1/2 inch tape and up to 84 inches or more.

Often a single width of material works well, but you may find that you will need a lot of tape, or narrow strips of fabric, for use in reinforcing corners where two panels intersect, or to build up material for extra strength at a panel edge or cut-out. These can be a nuisance to cut, and you should consider buying tape that is already woven to the widths you will need.

Various widths of tape are usually available at fiberglass supply shops that cater to people building or repairing glass boats, showers and trailers. Be sure that the material is treated with a finish that is intended for the resin type you are using. This is not normally a problem, as most of the materials sold in these outlets have a Volan A, or similar finish, and are suitable for both epoxies and vinyl esters.

Probably the most convenient feature that these tapes offer is the woven-in-place edge, or selvage, which makes the material stable and resistant to fraying, even in 1/2 inch widths. Widths up to 6 inches or so are usually stocked, and having the right width available will save time. This added convenience becomes even more important if you are using carbon fiber or Kevlar, as these materials fray and unravel even worse than fiberglass.

If you substitute for a specified weave, you should do two things to make sure the resulting structure will deliver the proper strength levels.

- Be sure that the fiber type is similar and the fabric weight is the same or higher in both the warp (length) and fill directions. If either is lower, use extra plies of material in an amount that will build up the same or higher weight of fiber in each direction.
- Perform a careful analysis before replacing a unidirectional weave with a tape of a different weave style.

WETTING OUT THE FABRIC

The primary objective in laying down the fabric and resin is to achieve a completely wetted layer of fabric without retaining extra resin that adds unnecessary weight. Most people who do this job first lay down resin and then put the dry fabric on top of the layer of resin. The fabric is then pushed down through the resin by a stippling action using a stiff brush. A squeegee can be used when resin viscosity is low, and there is a substantial amount of excess resin, so that you can flush out the air bubbles trapped under the cloth.

If the resin viscosity is simply too high to do a good job, look into an alternate resin, usually available from the same manufacturer, which offers the same laminate strength at a lower working viscosity.

It is a good idea to provide plenty of excess resin and work it out of the part so that the air bubbles are entirely removed, leaving as little excess resin as possible. Caution must be exercised however, when using the squeegee to eliminate excess resin in an open layup, as the layers of resin-wet fabric are a little springy. If too much zeal is directed at pushing out excess resin, the layers of wet fabric will compress under the blade of the squeegee and relax back to their normal position after the blade has passed. When they come up, air is drawn into the spaces between the yarns of the weave, badly affecting the final strength of the cured composite. Fortunately, one can see the trapped air under the surface of the wet fabric as a whitish blush. If you use a ply of plastic film between the squeegee and the fabric, this air entrapment is prevented, as well as the destructive effect of the squeegee on the fibers at the surface.

If you are just beginning to lay up composite assemblies, or if you have never noticed this effect in your own work, it would be a good idea to make an experimental layup in which you can see where this problem begins when using your personal techniques.

For those who plan to start a project using any epoxy system, it is also recommended that a copy of Mike Arnold's video cassette on construction of the World Record holding AR-5, *How It Is Made,* be purchased. It contains many good sequences on these steps with close-ups of both the carving and the "stippling" technique as well as many other useful construction details. This video is currently available from The Arnold Company to any purchaser, at the address shown in Appendix 1.

Another helpful reference is the one-page article written by Don Hewes of Light Aircraft Research Associates, P. O. Box 6394, Newport News, VA 23606, and printed in the September 1982 issue of *Sport Aviation.* Hewes discusses and offers tips on the handling of glass cloth. However be careful when using his method of laying out cloth on wax paper for pre-inserting. If you are using an epoxy system it will work fine, but the vinyl esters or polyesters will pick up the wax from the paper and make the next layer of structure impossible to attach. (The wax migrates to the surface and forms a release film of its own.) The pre-insertion method is fine, but get a roll of the same release paper that the adhesive makers use. It has a polyethylene surface on one side which will not transfer into your liquid resin. If that is too hard to find, PVC film, available at any dry goods or hardware store, also works well. One of the best is Nylon bagging film, sold by most of the supply houses.

VACUUM BAGS AND PLASTIC FILM BARRIERS

Finding the acceptable level of the final part's resin content can be a problem. Because of the need to keep the weight down, the importance of low resin content in the lay-up is repeatedly emphasized by all of the people who sell plans or kits. The effect of an over-eager effort to comply with the directions, while insuring a very light part, can also result in failure to achieve the desired strength level. One of the ways out of this trap is to use a vacuum bag or a thick plastic sheet coated with a release agent. (Some plastic sheet, such as polyethylene or polycarbonate, bonds so poorly that no release agent is required!)

If we seal off a space on the table or a tool with a shape on one of its surfaces that we want to reproduce, pump out all of the air from the plastic bag that surrounds it and let nature take its course, the air column above that tool will press on it with its entire weight. Because air is a gas, it pushes in all directions and the bag will push in on the tool's bottom and sides just as hard as it will on its top. If a hole or a tear in the bag develops, however, the surrounding air outside the bag will leak inside of it and the bag will no longer press on the tool.

This available pressure when the bag is sealed and evacuated is called the vacuum pressure. It will be as high as 14.7 psi at sea level on a standard day, and as little as 9 to 12 psi if you are at a high elevation, such as Denver, Colorado. Also, the vacuum pump will not provide a perfect vacuum, so the pressure will always be a pound or two lower than the theoretical maximum.

A common pressure gauge set up to measure the pressure inside the bag won't register, since it has a pin preventing the needle on the face of the gauge from going below zero. We therefore use a vacuum gauge, which is the same thing, except that it works backwards, with the needle showing a higher pressure as the pressure drops below room pressure. Since the weight of the column of air is only 14.7 pounds for a 1 square inch column, the maximum pressure the gauge can read is 14.7 psi. Sometimes pressure is calibrated in inches of Mercury instead of psi, and since a 1 square inch column of mercury that is 29.92 inches high weighs exactly 14.7 pounds, this gauge can never read higher than 29.92, on a standard day at sea level.

Atmospheric pressure varies a little with the weather and with your local elevation, so the actual maximum vacuum pressure obtainable in your shop may be a little lower or higher than these figures.

Vacuum pressure, as described above, is one of the standard methods used in molding composite parts. Using a plastic bag and an impervious (air-tight) mold or tool to which the bag is sealed, the part is laid up in between. **CAUTION: THE TOOL REALLY MUST BE AIR-TIGHT!** In addition to being perfectly air tight, the tool must also be dry and not enclose any liquids which will out-gas when the pressure is lowered. Ordinary fiberglass tools will pick up substantial amounts of moisture when stored outside, or even when stored inside, if the humidity is high. They must be dried our in an oven, or by being exposed to vacuum for an extended period, before use.

After the part is laid up, the bag must be carefully sealed around it, so that no air will leak in. A vacuum pump is then attached to a hose connected to the bag, and the air is drawn out of the bag. Mother nature applies pressure from the outside of the bag at whatever level of pressure has been drawn from inside, up to the maximum of about 14.7 psi. Although it doesn't sound high, a full vacuum will apply a pressure of more than one ton per square foot. Even when using a vacuum cleaner instead of a pump, the 3 or 4 psi it provides will yield 400 to 600 pounds per square foot! That is equal to a **lot** of sand bags!

Using the available pressure from the vacuum by means of enclosing the layup inside of a bag that is made of plastic film—usually nylon or pvc—allows us to squeegee out the excess resin, but allows no air back into the laminate! The laminate will also stay in the position where we pushed it while using the squeegee, which will give us a much more dense structure with greater strength than one made with the open layup technique.

When planning to suck up excess resin through the vacuum line, also plan on using a larger line so that it will not plug up with resin. A 3/4 inch diameter line should be large enough. Also, you must use a trap to accumulate the excess resin and prevent it from damaging the vacuum pump. A one gallon glass jug will work, as shown in Figure 7-1.

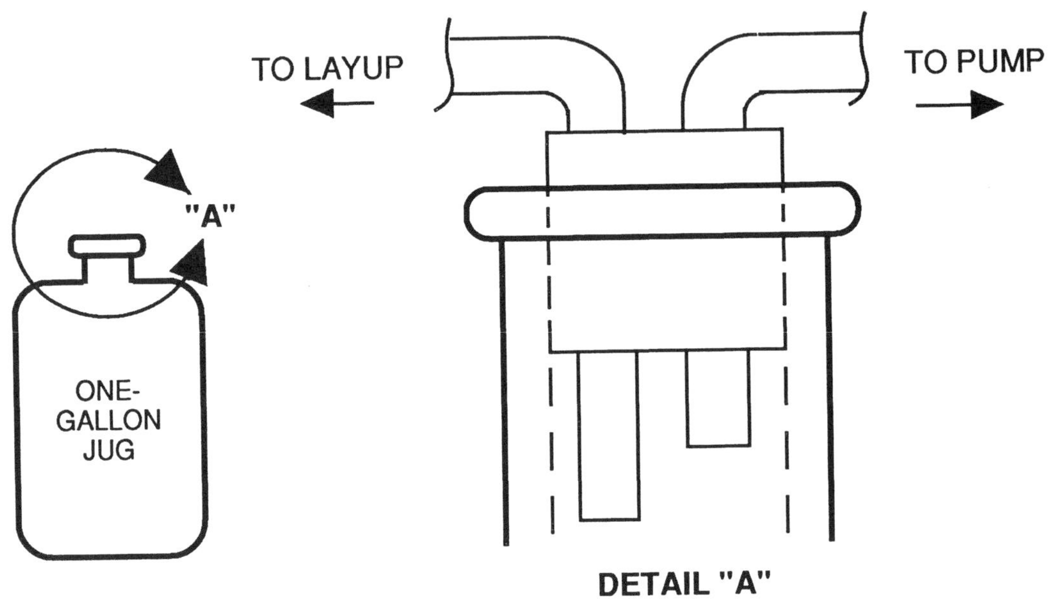

Figure 7-1. Detail of Hook-up for Gallon Jug to Accumulate Excess Resin

The final surface shape, smoothness and quality is heavily dependent upon the quality of the surface upon which the fabric is being laid. In the case of a hot-wire cut wing surface, any out-of-contour wobbles in the foam surface will be EXACTLY reproduced in the final outside surface, and will result in the need for a considerable amount of surface filling and sanding. Finish the original foam surface carefully, using enough final sanding and contour checking to ensure that an accurate reproduction of this surface will give a final contour that will require little surface change after the resin has cured. The strength of this part will also be greater, and its weight will be lower, as no sanding means there won't be cut-off structural fibers in the surface plies, and little surface filling will need to be added. For light-weight structures, which use only one or two plies of fabric, this technique is mandatory if the low weight desired is to be achieved.

A useful technique to insure a fair, smooth surface is to use an overlaid "surface former". This can be a piece of Plexiglas, Lexan, polypropylene, or polyethylene sheet, in any thickness of .050 inch up to as thick as 1/8 inch or more, depending on how much curvature the surface of the part being made actually has. In any case, the plastic sheet must conform to the shape of the part, which limits its use to flat panels and single-curved surfaces that have only slight curvature, such as a straight or constant-taper wing. The application of pressure to this plastic sheet may be made by means of an air-pillow and a large clamp, as shown in Figure 7-2. Such devices can apply only small pressure, however, and a vacuum bag, as suggested by Figure 7-3, works much better.

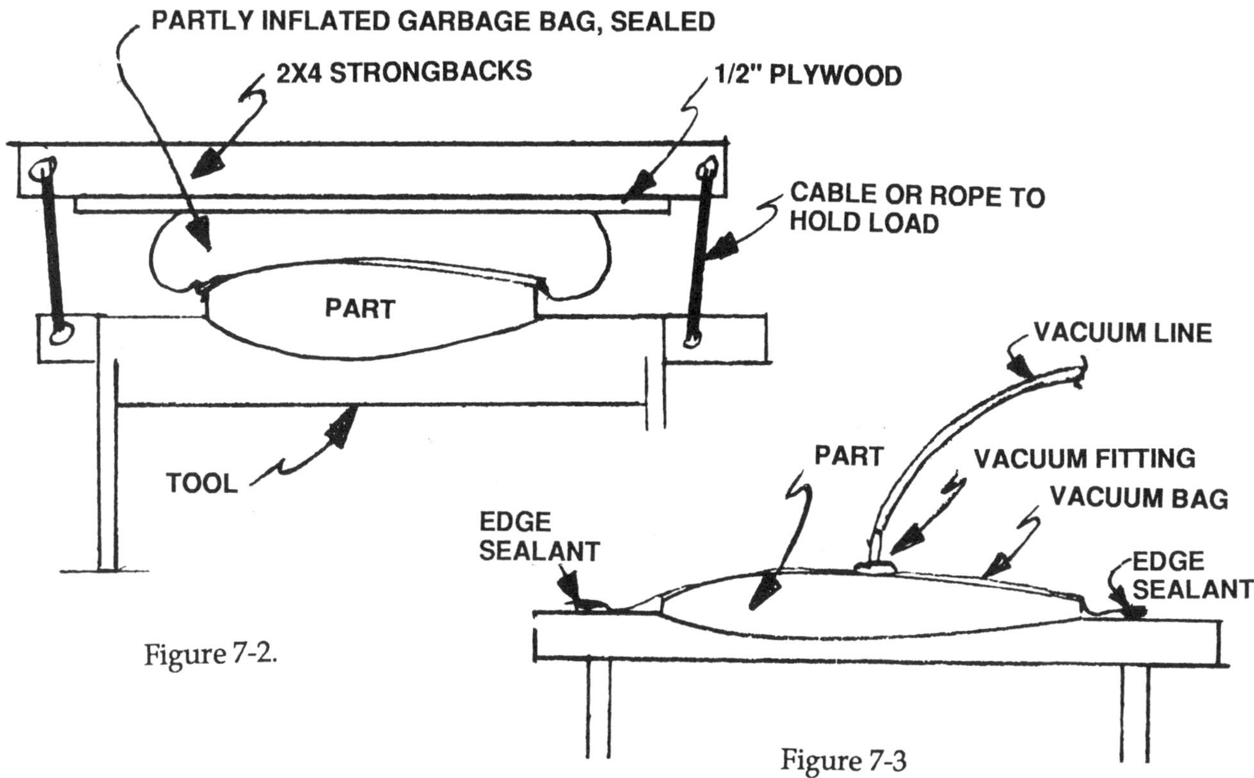

Figure 7-2.

Figure 7-3

RESIN CONTENT AND STRENGTH

The strongest structures that are made of glass fabric and resin will usually have less than 50% of their total weight in resin. The exact optimum percent of resin will vary with the style of fabric and the nature of the fiber orientation, ranging from 25% resin for a unidirectional layup made under pressure, to about 50%, or higher, for a cross-plied laminate using open weave fabric styles and made as an open lay-up with no pressure applied. These open weaves are commonly used in tooling and boat construction, and are necessary for producing hand layups with viscous resin systems, such as most epoxy formulations. With resin content of more than 50%, the structure of the laminate is almost always less than optimum. In other words, any resin that is more than 50% is excess weight. To check whether or not your laminate is close to optimum, weigh all of the fabric you will put into the laminate and then before trimming, weigh the final part after the resin has cured. For a good quality part the maximum weight should be only slightly more than double the weight of the fabric alone.

RESIN CONTENT VS. FIBER VOLUME

When resin content is discussed as a percent of the total laminate, the fiber is nearly always E-glass and the percentages quoted are by weight. An error is always introduced when one compares the resin content of different parts in which the resins used are not the same density, such as comparing a polyester layup to an epoxy layup. This error gets larger when different fibers are compared, such as comparing Kevlar with S-glass, which has almost twice the density of Kevlar.

When carbon fiber began to be commonly used, someone straightened this mess out by changing the term to "percent fiber volume", instead of resin content by weight. This lets us realistically compare the resin content of composites which use fibers of substantially different densities, without introducing any error. Unfortunately, the volume is difficult to measure accurately with home workshop equipment, so we usually use figures generated by one of the material suppliers or large aircraft companies. When glass fibers are involved, the resin content by weight can be measured at home, if you have a self-cleaning oven and a beam scale accurate to 0.1 gram. Just put your laminate in the oven for about 4 hours on the "clean" cycle, and weigh it before and after. Weight after, divided by weight before, times 100 equals the glass content in percent!

CUTTING AND TRIMMING

When using fabrics and resin, be sure you always have a good supply of razor blades and a large, sharp pair of scissors. The new circular razor blades that cut as a roller seem to work on glass and carbon fiber fabrics, but they are expensive. Have a setup immediately available for cleaning your tools after the job is complete. The epoxies can be nearly impossible to remove from the tools if they are allowed to sit overnight.

The tools mentioned above work well for glass or carbon/fiber fabrics and tapes, but not for Kevlar or Spectra 1000. When you use Kevlar or Spectra, it is best to start with a small sample of the fabrics, experimenting with the cutting tools until you are satisfied. If only a little of a lighter fabric, such as style 120, is to be used, it is possible to use Wiss #40 scissors, or one of the other "Big Bertha" models, such as those used by an upholstery or carpet shop. If you are going to do a lot of cutting, or if the heavier weaves of Kevlar will be cut on a routine bases, however, you should probably get a pair of the carbide-faced scissors, such as those sold by Pen Associates, 2639 West Robina Drive, Wilmington, Delaware 19808. (To see if you have a "light" or a "heavy" Kevlar fabric, simply check the thickness with a micrometer. Anything less than 0.006" thick is light.) Incidentally, this measurement is also the minimum finished, cured thickness per ply. If you are faced with a large Kevlar project, or have an arthritis problem, an investment in a pair of power-operated shears might be wise. The cost is about the same as 10 or 20 yards of fabric. This firm also has carbide razor blades that seem to work well on Kevlar, although they are rather expensive.

TRIMMING AND CUTTING CURED COMPOSITES

Cutting the final cured parts can be difficult or easy. The material you are cutting should be considered just as resistant to being cut as if it were steel or aluminum.

For example, a hacksaw works well on a finished glass laminate, as does a bandsaw with fine teeth and a metal-cutting drill bit.

Glass and resin have an abrasive effect on the cutting tools and carbide cutting edges last longer. Also, an electric jigsaw will work well, but you should have an abrasive blade for glass laminates and a special "both ways" tooth form for Kevlar. Often with Kevlar, material will not cut cleanly, but leaves a fiber fuzz on the cut surface. With glass laminates the cut is cleaner if you fully cure and post-cure the resin before cutting. Sometimes it also helps to increase the blade speed when possible. Even the special tools one obtains for cutting Kevlar will leave a fuzzy surface on the cut, and in most cases a wet sanding step is needed if the appearance of the cut is important. Do not attempt to dry-sand a Kevlar laminate. The fuzz just grows longer as you sand away the resin and leave more fiber exposed.

Where a Kevlar surface ply is to be used on an airfoil surface that must be subsequently filled, sanded and trued, add another ply of light glass fabric to provide something to sand off at the high spots without cutting into the Kevlar. To best take advantage of the Kevlar surface as an airfoil, the part is usually made in a female mold, so that little no finishing is required to achieve the airfoil surface-contour tolerance and finish.

MOLDING TOOLS FOR FIBER AND EPOXY

In order to make an assembly of materials that do not themselves have any fixed form or position, we must use a tool to locate and define the shape that the materials are to assume once fully cured. In the world of tool engineering there are many items that are called by many names, all of which help locate or define some portion of a finished part. Depending upon what job the device is called upon to perform, it may be referred to as a jig, a fixture, a pattern, a template, a model, a plug or any of a hundred other trade terms.

Each of these is a tool, in that it is a structure that is not a part of the product, but is used along the way to determine the shape, or some other attribute that we want the final product to have. In the case of composites and sandwich structures, the most commonly employed tools are female molds, bonding fixtures, trim templates and drill jigs. In the midst of all this is a major exception, Burt Rutan's "Moldless Construction", which avoids most of the tooling, but introduces its own problems as a consequence.

LOCATING THE SURFACE

The purpose of most tools is to fix the position of the outer surface of a part in construction from a layup of loose, floppy fibers that are wetted out with a somewhat runny liquid resin.

Since this surface often has a primarily aerodynamic or hydrodynamic function, such as a wing of an aircraft or the hull of a boat, the tool is normally used to define both the overall shape of the part and the contour and texture of the most important surface. Because the composite structure has little or no stiffness until the matrix resin sets up or cures, and because most of the critical part surfaces are exterior surfaces, the mold is shaped and positioned so that this critical surface will be forced against the mold face as the part is made. In an open layup, it is so forced by gravity as the liquid and fibrous parts of the layup settle into position and gel. In a vacuum bag layup, the very same thing occurs, except that vacuum pressure is the driving force. Such a tool is called a "female mold" because it has a concave form that matches the exterior of the part we intend to make.

If the part has an interior shape that is critical, we use a male mold, often called a "plug", on which the exterior is finished to the exact contour and surface finish that we want to duplicate on the molded part. We then lay up a part over this plug, and the inside surface of the part copies the shape and surface finish of the plug. (This is exactly the way in which female molds are normally produced.) If all this sounds elementary and simple, it is. The complexities and problems come when we implement this basic idea and actually make tools and build parts.

RUTAN'S MOLDLESS CONSTRUCTION

This is one of those simple and straightforward "why didn't I think of that", solutions to some rather complex tooling problems. The method consists of making a model of the part you want by using a suitable sandwich core material at full scale, and then covering it on both sides with a fiber-and-resin combination. This combination, when cured, will serve as a suitable sandwich-facing material. When completed, you have a finished sandwich structure.

Using this method, you avoid the time and effort needed if a plug and then a mold were produced first, and then the part was made using the mold. However, you must now smooth and finish the outside surface of the final part. This task can be rather time consuming, particularly when not enough attention has been paid to achieving perfection in the foam contours and the fabric conformity. When you begin, it is hard to appreciate how much work is involved in filling and sanding the final surface in order to arrive at the desired finish and aerodynamic conformity. Avoiding this labor can make the molding process more attractive.

THE SIMPLEST MOLD

Any surface that is used as a reference against which to cure a resin surface is a mold, but in practice, the flat surface of a press or an open sheet of aluminum or plate glass is used most often. Sometimes even the top of your work table, after being prepared with a mold release or a sheet of plastic film, can function as a molding tool for the production of flat panels.

The main difference between using a press and a flat plate as a layup surface is that a press permits the production of a smooth, flat surface on **both** sides of the panel. Both sides take on the smooth and unblemished appearance of the surface of the press platen. Also the press platens move together as the part is made, squeezing out the excess resin and achieving the best available glass-to-resin ratio. Every detail of the mold surface will be reproduced on the surface of the part, including all of the dents, nicks and dust particles. If you use your table top as a mold for producing flat panels, don't leave any wrinkles in the plastic release film that you are using as a mold surface, or they will be faithfully reproduced on the otherwise perfect panel you have made.

One neat trick in making flat panels, either in a press or on an open-plate mold, is to provide an edge mold, or "dam", by temporarily sticking an angle or square bar (or a flexible rubber bar for curved edges) to the surface of the plate at the point you want the edge of the panel to end up. Mold the panel up to the edge, as shown in Figure 7-4. You must also provide the edge mold with a mold-release surface, the same as on the surface of the flat mold, so that the part won't stick to it.

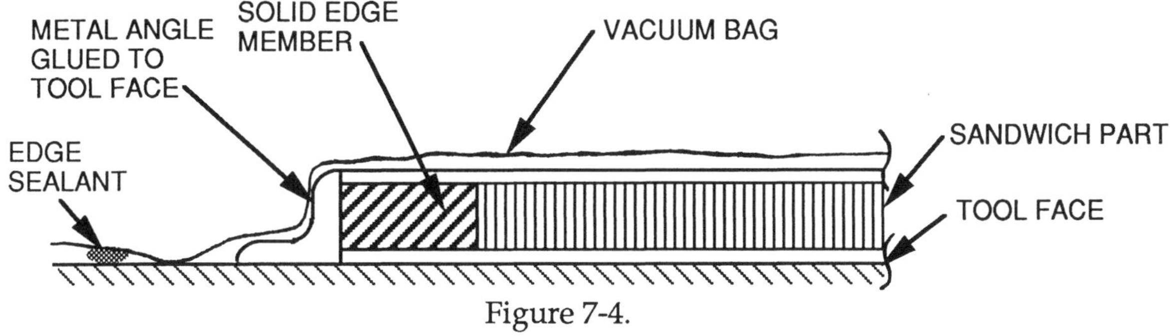

Figure 7-4.

With a simple set of edge-mold pieces, you can mold a flat sandwich panel that has a solid insert around its edge for later use of rivets or bolts for attachment. The panel then needs no edge trimming or shaping after it is cured. This can be a particular help when using Kevlar facings, as the Kevlar fabric is easier to cut as a raw fabric than it is as a finished panel. Another trick is to put a scribe mark in the surface of the tool where you will need an accurately located line on the parts for subsequent assembly operation. The scribed line will then be reproduced on the molded panel, but as a raised line rather than an indented line.

The same system can also be used on any tool when an accurately located trim line will be needed. The line is simply scribed in the tool surface, and then shows up at exactly the same place on each part that is made from that tool. Hole locations can also be set up on the tool this way, although it is better for holes to be marked with a raised spot on the tool, because the resulting dimple in the part can then be used as a centering hole for the drill point.

A TYPICAL FEMALE MOLD

To make a female mold, start with a master or plug of the exact shape and surface finish that is desired on the molded part. Plugs can be made of solid wood, foam or clay, or they can be made of an armature or frame, whose covered surface is foam, clay, Bondo, or nearly anything that can be formed to the desired shape, holding it long enough to make a "splash".

The splash is a wet resin surface that is backed up with glass fibers and a full-scale composite structure, so that it becomes the mold after it is cured. The exact resins that are used, and the structure backing up the resin surface, determine the stability and durability of the mold. When general purpose (i.e., garden variety) polyester resin is used to make a mold, the process is about as low-cost and fast as any known mold-making method. However, this system produces a mold that is a slightly different shape than the master because the resin shrinks a little upon curing. In addition, the cured resin continually changes shape with the passage of time, and with changes in temperature and humidity, so that the individual parts that are taken from the tool are a different shape, depending upon when a given part was molded. As a result, polyester resins are used only for tooling when there isn't a penalty for this slight change in the part shape, or when the mold cost is a major factor.

The most usual material used for molds is one of the many specially formulated epoxy systems in which the surface of the tool is made of "surfacing resin". The surfacing resin is the first part of the splash as the tool is started. The backup structure is made of glass or carbon fiber, saturated with another version of the same resin. The resin is formulated to give the best combination of workability during tool fabrication and dimensional stability after cure is completed. Lifetime of the tool should be many years when using any of the good tooling resins. Again, when fashioning this layup on the plug, the plug must be properly coated with a mold release so that the tool surface will not be distorted or broken when you later separate the tool from the master. You don't want to injure the master when separating it from the fully cured tool, because you may want to use it again to make another identical tool.

Before finishing the tool layup, it is usually necessary to create a backup structure in the form of a table, complete with legs, or a box with a set of milled-off pads, so that it rests squarely on the surface on which it will subsequently be used. The system of adding pads and milling them off to an accurate dimension is used extensively in both the automotive and aircraft industries, as this permits two or more tools to be placed next to each other, to form a surface that is much larger than any one tool by itself.

In the auto industry an entire reference system of "duplicate die models" is produced in this manner, so that any plant in any location may subsequently make hard tooling (steel or cast iron) and produce metal body parts for any model at any time. The usual aircraft mold has a set of legs that enable it to stand on the floor by itself and be used to lay up parts.

A QUICK METHOD TO PRODUCE LARGE SINGLE-CURVE MOLDS FOR A WING SKIN OR SINGLE CURVATURE DECK

When the part to be made is a flat plate, or a simple curved shape made entirely of straight-line elements, a piece of sheet metal may be used as the tool. Be sure, however, that no rivets or splices are located within the area to be vacuum-bagged, as any discontinuity in the surface of the sheet metal usually results in a vacuum leak in the final tool.

Tools for tapered wing skins are easily made by using thin sheet metal cradled in a grid of plywood or particle board ribs that define the shape of the mold. This was the method used for making the wing skins for the Voyager, at Scaled Composites.

The problem of overall mold rigidity is handled by making a base plate of the same thin sheet metal and gluing it to the base of the cradle assembly. The problem of smooth and wave-free contours from thin metal supported only by a rib every couple of feet is solved by making the mold surface out of two plies of the sheet metal, separated by a layer of epoxy resin applied with a linoleum paste spreader (a sort of saw-toothed spatula). This creates a sandwich surface for the mold. In every case, however, be sure that there are no splices or holes in the layer of the skin next to the part. If you will use an oven to cure your part at an elevated temperature, be sure to use one of the high-temperature adhesives to glue your tool together. Adhesives such as Hysol 934, will cure at room temperature, just like Safe-T-Poxy or West System resins, but can then be used for elevated temperature service, even at exposures as high as 450 degrees F. For fully contoured shapes that must be made of a glass layup, a tooling resin such as REN RP4004 will achieve the same remarkable performance.

Another large curved mold has been used extensively for production of curved decks having a variable degree of crown. An 8-foot by 12-foot aluminum plate of 1/2-inch thickness was mounted on a rigid frame, with only the two 12-foot edges held down. The centerline of the plate midway between these edges is then raised by a jack between the plate and the frame near each end. The amount of crown in the deck panel to be molded is then produced by raising the centerline of the plate by as much as four inches. If the upper surface of the deck needs to be molded against the plate, turn buckles are used instead of jacks, and the center of the plate is drawn down to form a concave surface.

THE TYPICAL MALE MOLD

When producing only one part, or when the complexity of going through several steps to make a female mold is otherwise not justified, most parts can be produced from a male mold even if the part has an outer-surface requirement for conformity and smoothness. An engine cowling, for example, can be made either from a vaguely similar cowling that is modified by the addition of foam and Bondo, or it can be made from a plug that starts with the engine to be cowled. It is mounted to the firewall and built up with bits and pieces of foam, fabric, clay and Bondo to achieve the desired shape.

In either case, the location of the joints must be carefully planned, so that the final part can be removed from the plug. In making a part from a male mold, the mold must be undersized by the exact amount of the layup thickness, so that the exterior surface has the correct dimensions. Because the surface must be sanded and finished, an extra ply of glass mat or fabric is often provided, resulting in an entire ply, which can be removed by sanding without dropping below the planned strength level.

When using Kevlar or carbon fiber as the reinforcing material, a surface ply of glass also makes finishing less exasperating, although it adds to weight without lending very much strength. If no extra ply is used, apply plenty of filler to the surface, using materials such as Featherfill, Bondo or flox. Then spline in the surface with a flexible straightedge, which fills all the low spots, rather than sanding off the high spots.

Depending upon the smoothness and conformity of the laid-up surface before filling, this may be either lighter or heavier than partly sanding off an extra ply. Where a gross amount of filler must be applied, such as a spar cap area in which the required number of plies of uni don't fill the cavity that is provided in the foam, it is sometimes better to glue on a thin slice of foam in order to provide the needed excess material to sand off. This results in a lighter part, because most of the conventional surface filler materials are between 25 and 50 pounds per cubic foot in density. This compares to only four to twelve pounds for the foams. In addition, the foams are easier to sand off than either the hard filler materials or solid glass laminate.

FINISHING A PART FROM A MALE MOLD

Whatever method you use to mold the part, certain requirements must always be met in the finishing operation: seal with a coat of resin all cut fiber ends that have been trimmed or sanded. Sealing a sanded surface can add many years to the life of your structure. Coat the exterior of the part with a barrier, usually paint, to prevent ultra-violet light from penetrating the structure. As mentioned earlier, all of the resin matrix systems slowly decompose under exposure to UV, and it is imperative that the UV be prevented from reaching the matrix resin.

UV barriers or inhibitors mixed into the resins only slow the deterioration of the resin structure, and even fluorescent lights give off enough UV radiation to cause measurable decomposition over a period of time as short as a year. An outside barrier of a good pigmented paint, however, may enable you to leave the structure outdoors permanently, with extremely small reductions in strength over 10 to 20 years. Conversely, the rate of degradation of bare structure that is made with uninhibited resin and left out in the sun can be fast enough to show a measurable reduction in strength in only a few weeks.

INFLATABLE PLUG MOLDS

You can avoid the finishing problem that is associated with parts made on male molds by using a female mold. Sometimes a female mold is too small for a particular part, preventing an operator from making the layup inside the cavity. Parts like horizontal stabilizers are usually made from two halves, even though the final shape would be under better control if the part were made as a single piece. Laying up the part over a male plug while using vacuum bag material as the layer next to the plug, may be a solution. After the layup is complete, the assembly is then slipped into the female mold and the bag edges are taped to the female. The bag is then evacuated, and the part is molded against the surface of the female mold as shown in Figure 7-5.

ADDITIONAL INFORMATION

More information on making composite tools is available in a 48-page *Epoxy Plastic Tooling Manual*, which is available without charge form Ren Plastics, a division of the Ciba-Geigy Corp., 4917 Dawn Ave., East Lansing, Michigan 48823. This booklet is a step-by-step, practical publication, and is an absolute "must" for anyone planning to make and use composite materials for tooling of any type.

VACUUM BAG MOLDING

The tools used in vacuum-bag molding are much the same as those used for open-layup molding, with a few special requirements. Some of these needs are obvious and some are not, so let's review them.

1. The mold must either be a vacuum-tight structure, or the bag must entirely surround the tool. The easiest option is to have a tight tool structure, and work on only one surface.
2. The mold must have a flat, smooth area surrounding the tool cavity in which the part is to be laid up, so that a bag material can be stuck down to the tool periphery, sealing off the part and allowing a vacuum to be drawn on the layup.

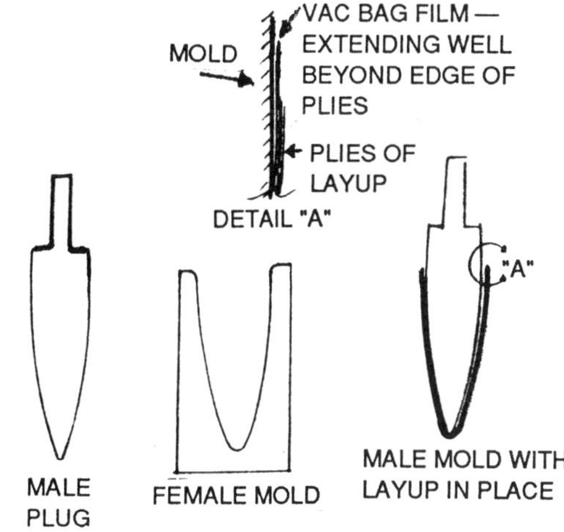

Fig. 7-5.

3. The tool must have a non-stick surface, so that the part does not become a permanent part of the tool-face following the cure cycle. Naturally, this is also true for open layup tools. This non-stick surface, usually called the release, can be built into the tool surface by providing a cured-on surface of teflon, or it may be provided by using various waxes, sprays, liquids, pastes, or even a thin plastic film like Saran, PVA or cellophane.

   Each resin system's ability to penetrate the release agent and stick to the tool is different. It is wise to try the system first on a small test part before you commit to making a large part in a new tool. Also, some of the so-called "permanent" release surfaces wear out after a few cycles, and somewhere along the way a part can become firmly bonded to the tool, even after the release system is thought to be working well.

4. In the case of plastic tools, some of the laminating resins attack the resins that are used to make the tools, with the result that making a part tends to destroy the tool. To escape this dilemma, spray a "gel coat" into the tool cavity, and lay up the part on the gel coat. When the proper amount of time is allowed to elapse between spraying on the gel coat and starting the layup, an excellent bond between the gel coat and the part results. The gel coat prevents the resin in the layup from attacking the structure of the tool. Since these gel coats may be obtained in a variety of beautiful colors (most fiberglass boats have an exterior surface formed this way), and also may be used to provide an excellent UV barrier, they are commonly used. The problem with their use in aircraft, however, is the added weight. Glasair designer Tom Hamilton calculates that the weight of the gel coat on the original Glasair is about 40 pounds or more for the entire airplane. All of the above information assumes that the tool is made of a composite material.

   When we make the tool of a metal, no such problems occur, and the use of a gel coat may be decided upon the basis of cosmetics and UV protection. The larger airframe companies are starting to use elecroplated nickel molding tools, and I suspect we will soon see them in common use as production tools. Cost and production method, however, effectively prevent their use by an individual builder.

5. Sheet-metal tools can easily be formed to mold parts when only simple curves are involved. Use epoxy to make the backup structure, even when using aluminum ribs and stringers behind the tool face. Any rivet will provide a continuing and totally miserable source of vacuum leaks into the bagged assembly during cure. They are nearly impossible to seal after the tool is made, because the back side usually includes a little misfit between the flange and the tool face. Even extensive sealing along fitted surfaces on the rear side will leave a few persistent sources for leakage. If you must use rivets, sink them in epoxy or sealant as you drive them, so that you at least have a fighting chance to seal the tool face.
6. Because the tool is impervious to air, and the bag is sealed to the tool face, the loads that result from the pressure of the bag against the part do not distort the tool. Even so, it is a mistake to assume that the tool structure can be flimsy, because tool distortion will occur any time the tool is set up on an uneven floor or table top, or when you lean on it as you lay up the part. Due to the need for tool stiffness, the resins and construction methods that are used tend to maximize the modulus (stiffness) and the stability of the cured laminates used to make tools, but do not emphasize strength. The epoxy systems normally sold and used to make tooling do this job well, and most of them provide strength levels far higher than needed.

VACUUM BAGGING MATERIALS

The use of bag molding is so widespread that a number of supply companies have set themselves up to sell the many materials used in the process. Such companies as Hawkeye Enterprises, Airtech International, and Richmond (all in the Los Angeles area), and Taconic Process Materials in Santa Maria, handle a broad line of films, release fabrics, sprays, liquids and waxes, breather and bleeder plies, dam stock, zinc-chromate tape (used for sticking the bag down to the tool face) and many other specialty items. Figure 7-6 shows a typical layup used in the high-temperature autoclave cure of a complex carbon fiber-based part, with a nearly endless array of special-purpose materials. Fortunately, the simple room-temperature cure requirement of most individual builders avoids the need for using many of these materials. The items you will need can usually be obtained locally and at modest cost.

BAG FILM, for example, may be made of any flexible plastic film that is air-tight and won't dissolve in the resin. One of the best is the plasticized PVC film sold at most dry-goods stores for use in covering the mattress on a baby crib. Also, the moisture-barrier film that is sold at hardware stores, or the plastic drop-cloths that are sold at most paint stores are usable for room-temperature cures. "Visqueen" is a common trade name. Be sure to avoid the very thin plastic films, as they usually have many small pin-holes which will make for many leaks in the final layup.

SEALING TAPE can be any double-faced self-stick tape, as long as care is used in keeping out wrinkles where the tape touches the film. An even better bet is ordinary caulking compound (the cheap stuff), sold in tubes for use in a hand-operated caulking tool, available at hardware and builders supply stores.

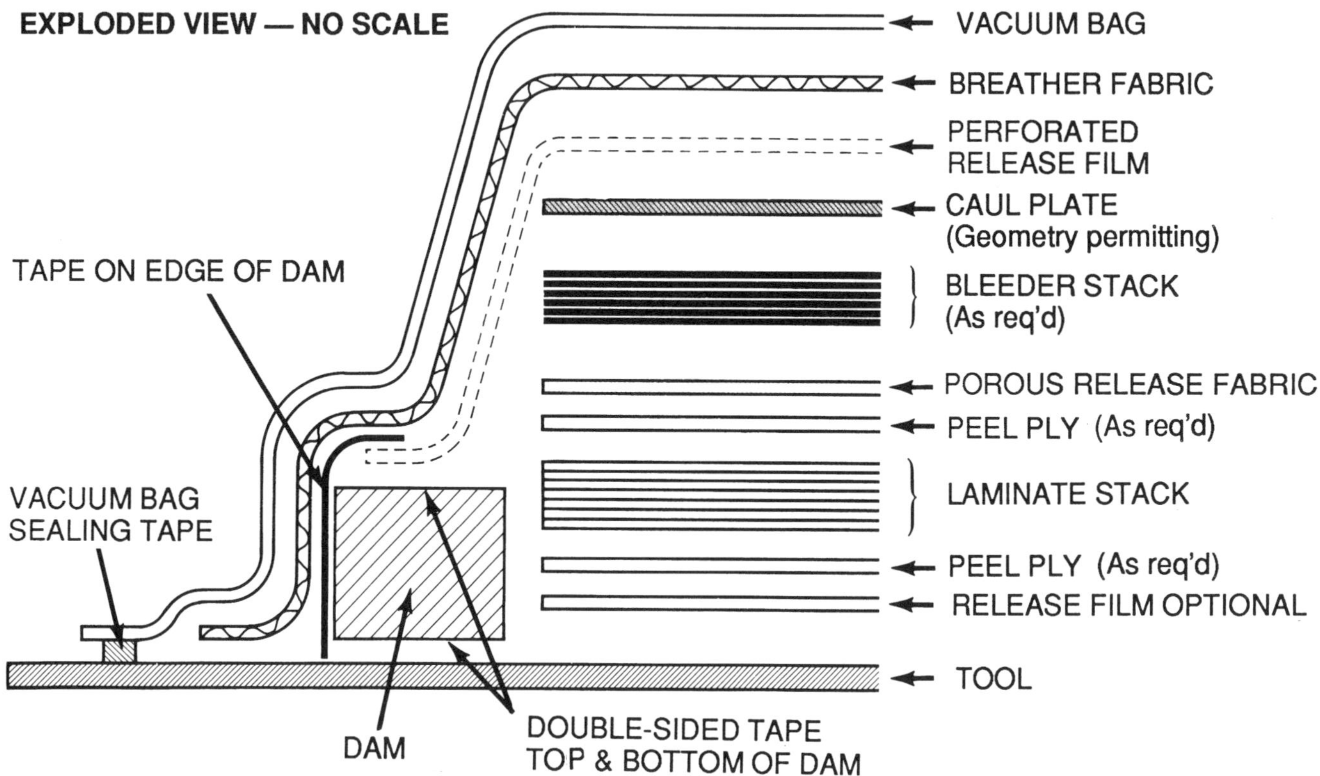

Figure 7-6. Typical layup used for a complex carbon fiber-based part.

RELEASE COMPOUND can be any good floor wax or paste wax, particularly those intended for automobiles or boats. Be sure, however, that it completely covers the mold and that it is buffed and polished. Often, layup instructions will call for as many as four coats, each buffed and polished prior to application of the next coat.

RELEASE FILM may be made of cellophane or of any of the same films as the bag. If you are using it between the part and a breather or bleeder ply, punch a zillion small holes in it, or it will seal off areas of the part and prevent the vacuum from covering the whole part. Also, check the film with your resin system to be sure it can be removed easily from the cured resin surface. (Cellophane tends to stick like crazy to some resins.) Commercially available perforated release films save a lot of time, and usually work better than the homemade variety.

A TEAR PLY is sometimes used between the layup and the bleeder ply, so that the resin soaked bleeder plies can be simply torn off the finished part after cure. Again, this material can be a nylon shirting material, or, better yet, one of the several fabrics made for this purpose and sold by the supply houses.

BREATHER PLY can be any fabric you are willing to throw away. Use enough of it so that air can freely pass along the surface of the entire part inside the bag.

BLEEDER MATERIAL can be the same as the breather, but you will need a lot more so that it can soak up the excess resin as it comes through the perforated release ply or tear ply. The thick, batting material sold by the commercial suppliers can be approximated by using the synthetic quilting that is sold at dry-goods stores. Be sure that it hasn't been treated with a non-stain material, as this will prevent it from soaking up resin.

THE VACUUM CONNECTION can be a length of copper or aluminum tubing going through the edge seal of the bag. Be sure to use plenty of sealant around the tube, as this is a common source of bag leakage. A better method of attaching the vacuum line is to use a flanged fitting, like that shown in Figure 7-7.

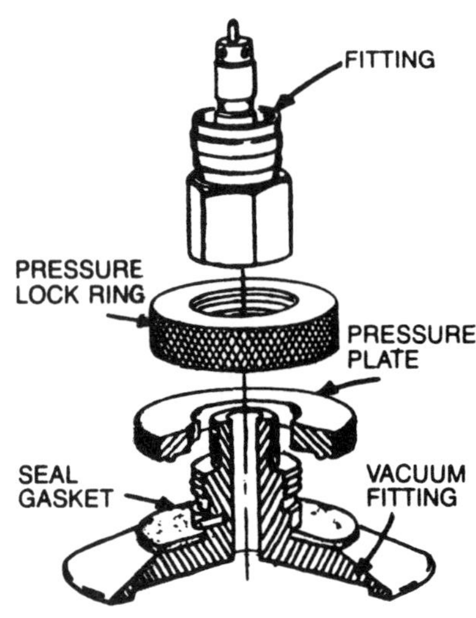

Figure 7-7. Vacuum fitting assembly (thru bag)
(Courtesy Hawkeye Enterprises)

In all of these examples of substitutions for readily available material that the industry would get from one of the specialty suppliers, be sure you test the material before using it. Sometimes the material that initially seems workable turns out to have some surface treatment or missing ingredient that causes problems in actual usage. For example, Hawkeye offers five or six different versions of the same material for some applications because of the small differences between resin systems.

One last tip on setting up a vacuum molding operation:

Allow plenty of excess material in the bag as it is being sealed in place over the part. About 6 inches folded into a 3-inch flap every three feet or so around the entire part is about right.

Figure 7-8 shows the way the bag should look. Even when using zinc chromate tape in several layers, the out-the-edge vacuum tap often leaks, especially after you bump it a couple of times while walking around the layup. You will be well advised to buy one of the special "through-the-bag" fittings designed for this purpose.

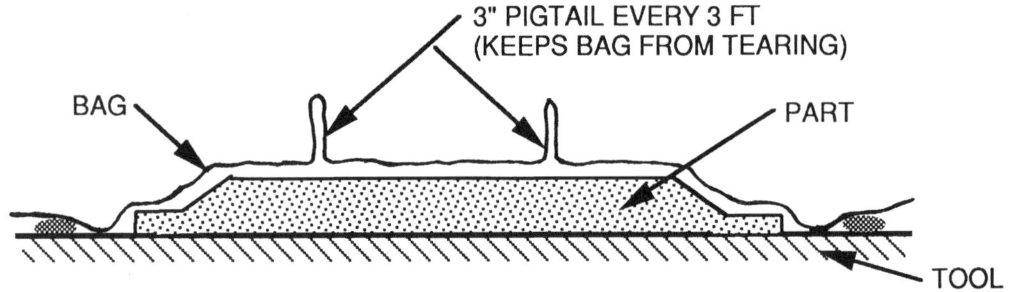

Figure 7-8. Typical bagged assembly.

CHECKING THE LAYUP

When all of the fittings are attached to the bag, and the part is ready for cure, check for vacuum tightness. If you have installed the extra vacuum gauge, this check is made by simply pulling a vacuum with the pump, closing the shutoff valve in the vacuum line, turning off the pump and observing the loss of vacuum with time in the layup. A well-sealed system will show a negligible loss of vacuum over a five or ten minute period. A poorly sealed bag (or a bad leak in the tool) will show a loss of about half the vacuum in as little as a minute.

For large parts, those having one or both of the plan dimensions more than three or four feet, it is a good idea to use more than one vacuum gage on the layup. This will show if all portions of the part are holding the vacuum, or if a leak in the tool is letting air into the layup at some unsuspected location.

A handy device to have when tracking down the vacuum leaks in your layup is a stethoscope. This doctor's listening device can easily and quickly pinpoint a leak, even with a fairly high noise level in your shop. Keep on looking for leaks, even if the resin has already started to set up, as lack of vacuum will always give an understrength part, compared to the part you will get under good continuous vacuum pressure.

TOOL LEAKS

One nasty source of bag leaks is not the bag at all, but the tool or mold, in which the part is laid up. It is obvious that a porous area in a fiberglass tool, or a crack in either a plaster or composite tool can leak outside air into the bag. That leak relieves the bag pressure in the area of the leak, and often will cause a defective part. These leaks are particularly hard to find when the layup is formed of a solid impervious core, and the facings are wet or viscous and tacky. In this situation, the area of the leak seals itself off from the rest of the part, and may be very hard to detect.

Be sure to check your tool with a blank layup (no part in the tool, but fully bagged as if there were one) to see if the tool is really sealed. If there is a problem with the tool, you can either repair the leak, or just bag the **entire tool**, backside and all. Remember that solvent or water picked up by the structure of the tool may out-gas under vacuum, and will act just like a leak with a very mysterious source.

HOW MUCH VACUUM DO YOU NEED?

The amount of vacuum pressure required to produce good parts will depend upon several things. The most important variable is the fit of the parts in the assembly before the bag is applied. If all of the pieces go into the mold without any appreciable misfit or warpage, only a few pounds per square inch of vacuum pressure will make good parts. Where the pieces in the assembly do not fit together well, particularly in tightly curved assemblies, or when bagging cores into female tooling, you will need much more pressure from your vacuum pump.

Make a dry layup on the mold(no resin), bag it with a transparent bag, and carefully inspect it under the actual pressure that your pump is providing. If the parts do not conform to the mold and to each other perfectly, you do not have enough pressure to make a good part. The pressure can be increased by using a better vacuum pump, but can only reach about 14.7 psi, even if you have a perfect vacuum and are at sea level. Practical shop maximum at sea level is about 13 psi, or 26 inches of mercury.

The easiest solution, if you already have a good vacuum source, is to make a better pre-fit assembly so that conformity to the tool is essentially perfect. Where this is not possible, the part must be made in an autoclave, which works just like a vacuum bag, except that additional pressure is provided, up to as high as 700 psi and more. As a poor second alternative, you can use an excess of a gap-filling resin, just like a hand layup. The part will then be heavier, but satisfactory parts can still be made, even with misfitting sub-assemblies under an inadequate vacuum pressure.

THE VACUUM PUMP

All of the above discussion assumes that a suitable source of vacuum is available, a convenience found in few home shops. In order to provide the vacuum, you can employ any of several devices. The handiest is your vacuum cleaner. If you use this method, be sure to allow sufficient bypass air (leakage), at the hose attachment, to permit adequate cooling of the motor, as the pump must run continuously during the **entire** cure cycle of the resin system, or about 4 to 10 hours for most room-temperature curing resins. If you use a standard four inch venturi from an old airplane's vacuum-driven gyro system, you will probably obtain both adequate cooling air and a substantially higher vacuum pressure than the unaided vacuum cleaner will provide. Either way, this source will only be sufficient for very limp or nearly flat assemblies, as it will provide 3 to 8 psi at best.

Another low-cost source is a device almost the same as an aircraft venturi, except that it is much smaller and fits on a garden hose, using water instead of air as the driving force. They are sold in hardware stores for use in draining flooded basements using city water pressure. Although it sounds terribly wasteful, they really do not use much water, as the venturi opening is quite small.

Most shops end up with a dedicated vacuum pump because they are so much more dependable. The cheapest pump is a junk air compressor. Even those in bad shape can work well, as the maximum pressure they will see in this service is only 14.7 psi, which is far less than the 50 to 100 psi that the typical compressor is expected to deliver. An excellent source is your local garbage dump, where fine compressors can be found in old refrigerators, discarded because the Freon system leaked. Any of these devices can be expected to deliver at least 10 to 12 psi of vacuum, enough for most purposes, and far more than the 3 to 5 psi of a vacuum cleaner. If you plan to get a first-class pump, they cost about $700 for the low-cost laboratory units, up to about $15,000 or higher for the units typically used in aircraft production shops.

If you plan to get a real vacuum pump, or even just an old compressor, you should also plan to plumb your shop with vacuum lines, so that you can locate the compressor far away from the working area. Most of them are very noisy, and represent a serious distraction if you must work close by. When running the vacuum line, locate several quick disconnect fittings at several appropriate locations in your shop, so that you can simply plug into the system when you wish to start the cure of another part. Remember to seal all of the joints in the vacuum line, as vacuum will find a very, very tiny leak in a joint. Even if your pump has sufficient capacity to allow such small leaks, the noise produced by a line leak can be terrible!

One additional unit is also a good idea. If you are to prevent the vacuum pump from constantly turning on and off every time you touch a hose connector, you should also include a plenum in the system. A perfectly good answer is an old compressed air tank from a shop compressed air system. The vacuum pressure load on the tank is much lower than the original load of compressed air that the tank was designed for, so it is still quite safe to use, even if it has a lot of rust and pits on the outside. If it actually has a hole in it, it still can be used, if you can plug the hole with some epoxy.

If you are interested in further exploring the typical vacuum bagging processes used by commercial shops, request a copy of a booklet, *Handbook of Layup and Bagging*, written by William McCracken, published by Mar Mac Engineering & Graphics and distributed without cost by National Aerospace Supply Co., at the address listed in Appendix 1.

RTM MOLDING

Although open layup and vacuum bag molding are probably the most commonly used molding methods, many other methods are also used. In the case of relatively simple shapes and a large number of pieces to be made, a method of closed die molding called, "Resin Transfer Molding", or "RTM", has found rather broad acceptance. When a vacuum is used to assist the flow of the resin through the dry reinforcing fibers, the method is called, "Vacuum Assisted Resin Transfer Molding", or "VARTM".

RTM is a process in which the part is contained in a matched closed mold, which has been filled up with a Preform which consists of all the reinforcing fibers intended to be contained in the finished part. This mold full of dry fiber is then infiltrated by a liquid resin which is pumped into the cavity full of dry fibers, displacing all of the air which was originally in the open spaces between the fibers. At the time of first use of the method, some fifty years ago, most of the parts were both small and of simple shape.

As the use of RTM became more widespread, much more complex and larger parts have proven to be quite successful in production situations. The largest part yet produced is believed to be the nose radome of the British-French Concorde supersonic jet transport, which was first flown in the 1960's. This part is more than sixteen feet long, and quite consistent in its control of critical dimensions.

In addition to success on larger parts, the method has also been quite successful on such complex parts as a complete thrust reverser assembly for the Boeing 757. Also, the blocker doors in the thrust reverser for all three engine installations have been made by this molding method. These parts were made with carbon fiber reinforcement, and then molded as a complete part in a single cycle. Many other large and complex parts can be found in current production.

Obviously several conditions must be met to be able to make successful parts using this method:

1. The value of the part to be produced must justify both the substantial mold cost and the often lengthy process development needed to assure success for any specific part.

Even though the part to be made may be quite similar to an existing production part, small differences can lead to the need to optimize the new and somewhat different process.

2. The closed mold tool must be sufficiently strong to resist the high loads imposed by internal pressure over the entire area of the mold, and it is not always obvious if vacuum assist will be needed, and if so, how many vacuum locations must be provided.

3. The preformed array of reinforcement fibers must be locked together by sewing, needle-punching, a small amount of compatible adhesive sizing, or some other method of preventing the fibers from changing their relative position with respect to each other.

4. The preform must fill the entire cavity, so that the moving liquid resin does not shift the location of the preform within the mold cavity.

5. The amount of net volume, not including resin, of the preform as a whole must be sufficient to result in at least 50 percent of the mold volume being occupied by fiber. Any percentage substantially lower than this figure will result in a relatively low part strength. 60 percent is a better goal, but this may be nearly impossible to reach while still allowing for easy passage of the liquid resin through the preform. In any case, the target should be within the 50 to 60 percent fiber volume range.

6. Air remaining within the cured molded part is a serious defect, and must be reduced to the smallest amount practicable. Commonly used steps to reach this objective include a vacuum assist, which removes existing air before entrance of the resin into the cavity, and also helps add to the injection pressure which is moving the resin through the preform. Vacuum can be particularly effective, since it adds to the pressure moving the resin, but does not add to the overall pressure load on the mold.

7. Resin to be used must be carefully chosen to allow for both a very low viscosity during the injection cycle as well as a sufficiently long time before early stages of the cure begin to increase the resin viscosity. These requirements usually result in a need for very closely controlled temperatures of both the resin and the mold during the injection process.

Even with all the above requirements, as well as the many more not covered in this basic discussion, RTM has become a useful and commonly used method for molding composite structures.

SCRIMP MOLDING

SCRIMP (Seemann Composite Resin Injection Molding Process, patented by a Mr. Seemann) is a relatively new version of RTM, in which the reinforcing material is held against the mold by a vacuum bag, the same as in normal vacuum bag molding. In this method, however, the layup is made up of dry fiber, in all the orientations and layers required by the finished part, just as in RTM molding. However, the resin is then introduced to the layup by only a vacuum in the bag, so that reasonably light molds may be used. The resin may then be polymerized either by moving the entire assembly into an oven, or by use of an ultra violet curing catalyst in the resin, so that a relatively fast cure can be accomplished after a reasonably long time of working with the resin in liquid form.

One of the important incidental benefits of the SCRIMP method is the ability to seal off the liquid resin from the workers producing the part, as compared to an open layup method, in which the fumes from the resin mix represent a health hazard to any individual working nearby.

SCRIMP, like RTM, does not necessarily make for a faster production cycle, but very large one-off structures, such as boat hulls of 60 to 100 feet in length, which can be produced using essentially the same mold which would be required for a simple hand layup. As in vacuum bag molding, however, the mold must be absolutely air (vacuum) tight, with no leaks whatever!

QUICKSTEP MOLDING

One of the newest production methods recently introduced, which is also capable of very high production rates, is the "Quick-Step" process. This is a novel method in which a set of very light matched metal molds is used, with an intermediate temperature for optimum resin flow, a curing temperature for final cure and a cold temperature for release of the molded part are all provided from separate reservoirs of fluid (either water or oil) which are pumped into the mold, each in turn, by a set of quick-acting valves. Molding pressure is provided by a separate system using either pumps or pressure chambers as a source, while very fast infiltration of the liquid resin and expulsion of excess resin is aided by physically jolting the molds by use of a fluid hammer in the pressure circuit.

Parts about two feet by three feet with a thickness of one inch have been produced by this process in only 30 minutes total cycle time. At this time the process is being exhaustively examined by both Boeing and the Australian government.

For further information, you should consult their web page, www.quickstep.com.au.

COMPRESSION MOLDING

One of the most common molding methods, compression molding, involves the use of matched metal molds which completely surround the fiber and resin, and are shaped so that the volume of material within the mold can be compressed as the mold halves are closed. Most any kind of fiber and resin can be used, provided that the resin has the appropriate flow, gel, shrinkage and release properties. One of the most common is a chopped, epoxy preimpregnated tow, in cut lengths of one-quarter of an inch up to an inch or more in length. In addition to cut lengths of chopped fiber, chopped up fabric prepregs are also often used, furnished as either bulk or sheet molding compounds.

Larger parts with thin wall sections are more difficult, but quite common. Most typical are small brackets, end fittings, stand-off spacers, as well as many specially tailored EMI and other electrically sensitive items.

Strength of the finished part depends upon the fiber and resin used, as well as upon the effect that movement of the resin during cure has on the direction of the reinforcing fibers within the finished part.

Because very high rates of production can be achieved, the method tends to be favored by automotive manufacturers, while only infrequently seen in aircraft or space applications. The exception, of course, is in such items as pulley wheels for control cables, and other such standardized items used in large numbers.

OTHER MOLDING METHODS

In addition to the production molding methods discussed above, other processes are everywhere. Only the imagination of the Tooling and Process Development Engineers can tell us what is next. Already common are the use of braided tubular preforms to make non-linear and non-uniform tubular structures, hybrid methods involving the use of filament wound tubes or braided tubes which are subsequently molded to the shapes necessary to make "H" beams, "J" spars or other such unusual shapes before final cure in a closed mold. Such subsequent operations performed prior to cure can also include insertion of metal or composite reinforcements for load points, and many others. A great many such ingeniously devised parts are already in production, and the number of those about to be invented is even larger.

Filament winding and pultrusion are two more ingenious production methods for producing large numbers of parts. Size of parts vary from a millimeter or two in diameter, on up to three or four feet for pultrusions and twenty feet or more for the larger filament wound structures.

A revealing description of many of the latest innovations can be found in the book, *Composite Aircraft Structures* by Michael C.Y. Niu, available from Technical Book Company in Los Angeles.

CHAPTER 8

WET RESIN AND HONEYCOMB CORES

The most efficient sandwich structures, those which have the highest strength-to-weight ratios, are made with honeycomb cores rather than one of the several foam or wood core materials used in most projects. The foam or wood cores make an excellent structure, but the honeycomb cores nearly always give more strength and stiffness at a lower weight. The difference may be only a few percent, but in some cases it is as much as a hundred percent or more. Refer to Chapter 3, Figures 3-2, 3-3 and 3-4. As a result of this well known fact, nearly all of the sandwich structures seen in both large transport and military aircraft use a honeycomb core, but seldom make use of any substantial amount of foam or wood core sandwich.

The most common honeycombs used in these applications are no longer one of the aluminum honeycombs, but in recent years, are more often one of the many types of Nomex honeycomb or its newest cousin, Korex. The aluminum cores still cost less than the Nomex or Korex honeycombs, but the clear superiority of Nomex or Korex in corrosion resistance, abuse resistance, high toughness and ease of assembly has made this family of materials a clear winner in the honeycomb sweepstakes. This superiority is particularly noteworthy when the sandwich weight is near the minimum that can be produced, and the most difficult requirement is that of resistance to service abuse, handling damage, normal service usage, and other such incidental loads. Since the total area of structure whose primary design loads are so low that they fall into this category is relatively large, particularly on the larger civil transport aircraft, more and more of the total area seems to end up as a sandwich having a Nomex core.

Paralleling this trend in core materials away from the "lowest-cost" candidates, and toward the "lightest weight/best service experience" cores, has been the emergence of a much more common use of glass fiber, carbon fiber or Kevlar facing materials. Even carbon fiber facing materials are now becoming quite common in the large transports. The increasing use of these fibers is due to the same reasons for the steady increase in Nomex usage. Many combinations of these materials are simply tougher and more damage-resistant than the equivalent aluminum sheet, and are substantially lighter at the same thickness. The strength of 2024-T3 aluminum is approximately the same as cross-plied E glass, but the glass laminate is thicker at the same total weight per square foot. Because the sandwich is often critical in edgewise compression or bending, where failure occurs by skin buckling, those comparable sandwich facings seem to be much stronger, failing at higher loads, when made of glass, Kevlar or carbon fiber than they are when made of aluminum.

In the case of Kevlar, the difference is startling, as the fiber is substantially lighter than glass fiber and has a higher tensile strength. Compressive strength is not quite as high as glass, but the sharply lower density increases the sheet buckling strength observed at the same skin weight.

Kevlar is sometimes compared to carbon fiber in its performance as a structure, but it is really a quite different material than either glass or carbon fiber. Both strength and modulus are different, but the real performance advantage shows up in those real-life structures that must be low in weight, and still make it past all of the bumps, cuts and scrapes which are inflicted on every part of a real airplane.

In addition to this remarkable toughness advantage which Kevlar exhibits over glass, carbon fiber and aluminum structures, two other startling differences are usually noted in structures made of Kevlar fibers. First is its refusal to fail under load in the brittle manner normally seen in glass and carbon fiber structures. It stretches, groans and then stretches some more, acting much like a ductile metal part instead of failing in the brittle manner of most other advanced composites. The second difference is the way a Kevlar structure simply "eats up" vibration! In spite of its high strength and modulus, it has the highest damping of any of the high-performance advanced composites in current use. The result is that sandwich structures using Kevlar facings on a core of Nomex honeycomb are becoming one of the aerospace industry standards, where minimum weight must be combined with superior toughness, high damping and durability.

For many years facings based on carbon fibers have been regarded as somewhat fragile, in spite of their sometimes very high strength. This is due to the fact that these fibers usually fail in a very brittle manner, compared to Kevlar, and sometimes, to glass. However, some of the newer carbon fibers, notably the T-650, T-850 and T-1000 products from Toray, exhibit more toughness, rather than less, compared to E glass, and almost as much as S glass. The number in the name for a Toray product designates the approximate fiber tensile strength in ksi, or thousands of pounds per square inch, indicating that the T-1000 will deliver around **one million pounds per square inch** in tensile strength. Not bad, for a yarn that costs only about $35.00 per pound!

Much of this performance depends upon the resin system used with the fiber, and, unfortunately, most of the wet resin systems available to home project builders do not develop the outstanding attributes of these newer fibers--at least, not to nearly the degree shown when some (but not all) of the newer toughened epoxies or polycyanate prepreg systems are used. Unless you are prepared to use prepregs and cure your parts at a substantially higher pressure than atmospheric pressure (a vacuum bag alone gives a maximum of only atmospheric pressure, even under ideal conditions), these newer, very high strength carbon fibers may not deliver the expected strength or toughness.

All of this is fine for those aircraft builders who have an autoclave or oven for elevated temperature curing of their composite structures, as well as cold-storage capacity to handle the various prepregs (fabrics pre-impregnated with a resin, ready to be cured). However, the individual or small shop desiring to achieve these same benefits in his composite structure has quite a different problem. Not only does the need for these expensive facilities stand in his way, but most prepreg makers are reluctant to sell to individuals in small quantities.

As a result of the many problems noted above, the combination of either glass, carbon fiber or Kevlar facings on any honeycomb core has been rare in home projects. One of the principal problems, of course, is that when wet resin is used instead of prepreg, the resin tends to run down the inside of the open honeycomb cells and add a substantial amount of excess weight.

In spite of such problems, a few pioneers have begun to use the combination of wet resin, woven fabric skins and Nomex honeycomb quite successfully, and there are now some answers that make the process a little more feasible. The methods they have used to solve the wet resin problems are outlined here, so that the individual builder can now produce parts similar to the high-efficiency, high-toughness and minimum-weight parts used on commercial airplanes.

THE MOLD

We must first have a mold upon which, or into which, the part can be assembled, or "laid up". Mold construction is covered in Chapter 7, and should be carefully reviewed if your mold is not already in existence. Even if you already have a mold used previously by someone else, be sure to inspect it carefully, filling in any scratches, gouges or dents, polishing the mold surface, and drying it thoroughly. Moisture in a mold is normally present if the structure has been stored where high humidity or exposure to rain or other moisture could affect it. Even if you must rig up a temporary oven, using cardboard and an electric heater with a fan, it is usually well worth the effort. Also, be sure to allow adequate time. At least four to eight hours is required for this entrapped moisture to escape.

PREPARING THE MOLD SURFACE

When you are satisfied that the mold is dry and has the correct shape and surface finish, you are ready to make a part. The first step is to thoroughly clean the mold surface so that no dust particles or bits of debris will mar the final appearance of the molded part. When the mold is absolutely clean, coat it with a mold release, such as a paste wax used for floors or auto body polishing, "Frekote" #44, or one of the many other available mold-release materials.

Coat the entire surface of the mold, and polish out the released surface to retain the surface finish quality you expect on your part. Remember that the surface of the final part will come out **exactly** like the surface of the mold. If you are using a material for mold release that is new to you, make a trial part in a test mold first. Each resin system reacts differently when used with any specific mold release, and you should test it out first, before committing an entire part.

As soon as the mold surface is ready for the first skin, put a dust cover of plastic film over the mold to keep it dust-free while you prepare the first facing ply. Obviously, the dust cover must also be entirely free of dust before you put it over the mold. As a matter of standard shop practice, most production shops lay up parts in an entirely different location from where the usual grinding, cutting, and other dust producing activities take place. If you are using the same room for both activities, be sure to complete a thorough cleaning of everything in the room, before beginning any layup operations. Also, remember to avoid any operations which might produce dust during the lay up process.

THE OUTSIDE FACING

Although we do not know if your part will have the outside of the finished part represented by the side that is next to the mold, this first ply of skin to go into a mold is usually called the "outside" ply, or the "mold-line" skin. This skin may be wetted out using either of two methods. The first and simplest is to lay the dry fabric into the mold cavity and pour an excess of resin in on top of it. The resin is then worked into the fabric by using a squeegee to push it back and forth until all of the fabric is entirely wetted out. At this point, use the squeegee to push the excess resin onto a blade and remove it from the layup.

One of the main problems with this method is that the mold side of the fabric is hidden so that it cannot be checked for complete wetout, and the exact amount of resin in the final squeegeed ply may be either excessive or inadequate. It is more common, therefore, to measure the area of fabric being put into the mold, weigh out the amount of resin needed to give the desired resin content, usually 50% to 55% of the total weight of the Kevlar and resin combined, or 40% to 45% of the total weight if glass fabric is used. Add this entire amount of resin, and then work it all in with a squeegee. The 55% figure is easily achieved by weighing the Kevlar and adding that weight, plus 10% more, in resin. Do not use the squeegee with too much vigor, as the fibers are quite fragile at this point, and can be easily broken or pushed out of line.

Because of the danger of tearing fibers in the woven fabric, particularly when using Kevlar, and the problem of inspecting (or not being able to inspect) the mold side of the fabric for wet-out, a better method is to use the pre impregnation method described above, but perform it on a clean table top, with the fabric placed over a ply of plastic release film. After the resin is thoroughly impregnated into the fabric, while it is in place on top of the plastic film, the entire piece of impregnated fabric and film is transferred to the mold cavity, being carefully carried by the plastic backing film. The squeegee, or a roller, is then used to work the layer of impregnated fabric into intimate contact with the surface of the mold while the plastic film is protecting it. The film is then removed, leaving the fabric and resin in place in the mold.

An additional benefit of this method is that several plies of fabric can be pre-impregnated at the same time on the table and moved into the mold at one time. Performing this operation at a work table instead of a mold cavity allows more careful alignment of the warp and fill directions of the plies of fabric, as well as easier inspection of the assembly for absence of wrinkles, voids, air bubbles and the like.

THE TIME FACTOR

If you are using one of the room-temperature curing resin systems, the time needed for assembly is quite limited, since we must place and assemble a core and another facing before we can get the vacuum bag into place and apply the proper pressure to the layup. There are two ways to solve the time problem. The first is to stop the layup after assembling the mold-side plies of fabric and the Nomex honeycomb, applying the bag at that point, and completing the cure of the partially built assembly. The second facing is then applied later, as an entirely separate operation.

A second solution is to use a slower-curing resin system, which will allow for the completion of all the steps in even the most complex assembly before you run out of time on the resin. Both epoxy and vinyl ester systems are available in these slow-curing versions, but their use is not widespread, and you must contact the resin supplier to make arrangements for obtaining the proper material.

The best method of solving the time problem is to use one of the many elevated-temperature curing systems available. These systems will allow several hours of layup time, and then cure in an hour or two when the temperature is increased to the proper level. Actual cure temperatures vary from one system to another, but typical values range from as low as 125 degrees F up to 350 degrees F or higher. If you are to cure at or above 200 degrees F, it becomes much more attractive to use one of the low cure temperature prepregs. These systems give an allowable layup time of at least a couple of days, up to as much as four weeks or more for some systems.

APPLYING THE HONEYCOMB

After the mold-side facing is in place, the honeycomb core is positioned in the layup. It is essential to have all of the edge shapes, cut-outs and inserts at hand and, preferably, edge-bonded into one single piece of core material. If you use this edge bonding step as a pre-assembly, be sure to align all of the parts of core bonded together, so that there are no steps in the thickness.

Inserting several different pieces into a layup can be time consuming, and may cause a mistake that will result in a defective part. When bonding several different pieces of core into a precurved core assembly, it is well to do this edge bonding under a vacuum bag in the mold so that the final core assembly will have the relatively rigid splice lines reinforcing the desired shape of the piece, instead of fighting it.

The low-density Nomex cores (less than 2.5 pcf) will form into the mold easily, under only a few psi vacuum pressure, without precurving, but any splices made while the core was in the flat position will stoutly resist bring formed to the shape of the mold. This will result in a flat spot in the finished part, or even an unbonded area in the core-to-face bond. The only sure method to avoid this problem is to use the bag molding splice technique mentioned above. This will ensure that the splice has been cured while the core was being held to the proper contour, and that the spliced assembly will fit exactly into the shape of the mold.

When positioning the honeycomb on the wet glass or Kevlar skin already in the mold, do not slide the core if it needs to be repositioned. Sliding it on the fabric will tear many fibers, and may also move the skin and cause a wrinkle. Simply lift the core up, move it to the new location, and press it back into place again. Wrinkles cured into the skin are a very serious structural defect, and should be avoided.

The low-density Nomex honeycomb cores will be quite flexible and floppy, particularly in large piece sizes. This leads to a somewhat ambiguous or ill-defined length and width, and the core assembly should be fitted, checked and then rechecked before cleaning the mold and starting the layup. The same characteristic that makes the material's length and width directions uncertain, however, also permits some stretching or compressing. A little practice will make the fitting-up job much easier than it is when using foam or wood cores.

Some of those who have made honeycomb sandwich structures using the methods described here feel that, if the maximum level of core-to-face bond strength is to be achieved, the honeycomb should be primed with a little adhesive. For example, Dr. Leo Windecker, in making the prototype "Avtek 400," primed all Nomex honeycomb by dipping the slices in a solution of his laminating resin (Dow 382 epoxy), thinned with 80% MEK solvent. He reasoned that this would give him the highest possible peel strength. However, several other builders making similar structures report that the peel strength is superb without the need for taking this extra step.

Because each resin system will behave slightly differently in this regard, it is recommended that you make some small test specimens using each method, and evaluate the bond strength that your process and materials actually deliver under your production conditions. If priming is necessary, the usual method is to roller coat each surface with the same resin which is being used in the facings.

Should the resin have too low a viscosity, resulting in resin running laterally down the facings that are in the sides of a large, deep part, a small amount of Cab-o-Sil (5% or so) may be added as a thickener, or thixotropic agent. This additive has little effect on the strength of nearly all resins, so long as the amounts used are within reason. On the second side skin, if the resin seems to be draining out of the facing plies and running into the open honeycomb cells, the same additive can be used to help minimize the problem.

ASSEMBLING THE SECOND-SIDE, OR BAG-SIDE SKIN

Prior to placing the second side skin on the honeycomb, it **must** be preimpregnated with the required amount of resin. The preferred method is to lay up the skin on a plastic film on a work table, the method described previously for the mold-side skin. If the facing is to be composed of more than one ply, all the plies should be made up as a single preimpregnated skin on the table, as the open cells of the honeycomb will effectively prevent the compacting of the plies together after they are in place on the core. **DO NOT** squeegee the surface of the second side skin after it is in place in the tool, as the action of the squeegee will drive resin out of the facing laminate and down into the open honeycomb cells. A hard-surfaced roller may be used to push the laminate into place, or it may be left in relatively poor contact with the exposed core face, since the vacuum bag will take care of this task quite nicely.

Where small pieces of the fabric facing material must be added separately to the layup, they may be pre-impregnated on a table with a foil backing, rather than the plastic film backing. This permits subsequent cutting and trimming of the piece to exact size, with the inelastic metal foil preventing any change in shape of the trimmed piece while it is being trimmed and smoothed into place in the layup. After placing the ply in its final location on the lay up, the foil is gently peeled off, leaving the fabric and resin perfectly placed. By using this method, relatively small and intricately shaped plies can be accurately located. (Dick Rutan, who told me about this trick, refers to it as the "Dave Ronnenberg" method.)

After completing the layup, bag the part as described in Chapter 7.

CHAPTER 9

DESIGNING COMPOSITE STRUCTURES

SOME APPROXIMATE METHODS TO DETERMINE STRESS AND PART SIZE

In the design of any complex structure, it is required that many specialties be brought to bear on an overall problem, and that the designer have the specific knowledge required in each of the special fields. For the purpose of this chapter, we will consider only the narrow subject of choosing and placing the materials in a design where it performs a specific structural function. Of course, it really is not possible to completely restrict the design function to such a narrow degree, and we must do some stress checking so that we can put the right amount of material into the part to properly carry the loads it will see in service. Also we will not attend to the earlier requirements of the design function in which the shape of the part is determined, even though the functional shape and the structural design sometimes interact with each other and cannot be entirely separated. For the moment, then, we will only address the choice of materials and discuss putting them together to fabricate a structure.

Let's review some of the fundamentals of composites which will affect our design decisions.

1. When we use fibers to carry loads, they will carry them only in the same direction as that in which the fibers are placed. The fibers that go in another direction, such as the crossing yarns in a woven fabric, serve only secondary purposes.

2. The facings in a sandwich structure will carry only loads that are in a direction that lies within the plane of the facing. Any load or component of a load that runs in another direction must be carried by the core, even though it must go through the facing to get there.

3. The fibers that are woven into a fabric and have the crimp which results from going over and under each other will not carry quite as much load as those that are laid parallel to each other without any crimp. This reduction in strength is always at least five or ten percent, and in some carbon fibers, can be as high as fifty percent.

4. The resin that carries the fibers is usually weak and soft compared to the tensile strength and stiffness of the fibers, and any excess resin beyond that which is actually necessary to fill the spaces between tightly packed fibers will result in a substantial reduction in the composite's strength.

5. Air bubbles trapped in the composite will not carry any load at all, and will seriously degrade the strength of the laminate or sandwich. Two or three percent of these voids can degrade strength as much as ten or twenty percent.

SPAR DESIGN

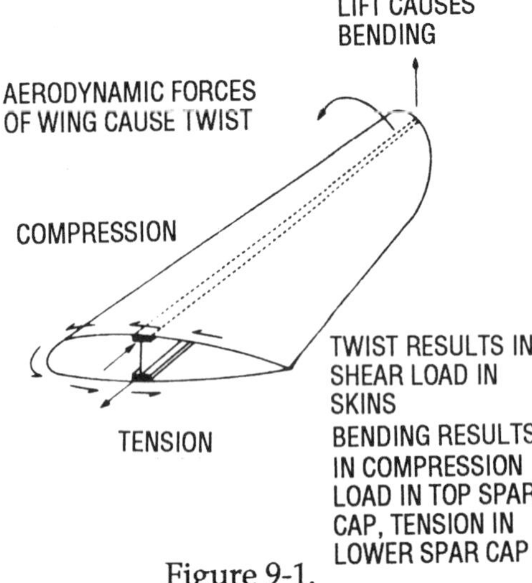

Figure 9-1.

In a composite wing design, the spar usually carries all of the wing bending loads, while the outside skin carries the wing twisting loads as shown in Figure 9- 1. Since the spar-cap loads almost all run in a single direction, we will choose a unidirectional glass fiber structure. To determine how big the cap must be, it is necessary to first approximate the stress analysis, which is necessary to set the sizes of the different pieces of the structure.

Let's assume that our airplane weighs 1000 pounds total, including people and fuel, must resist a 10-G load factor, has a 24 foot span, a wing section that is 9 inches (0.75 feet) thick at the root, and whose spar will be located at this 9 inch thick point. In order to calculate the load on the spar cap, we must assume a span distribution of the load, but for a first approximation, we will suppose that the half-weight of the airplane will be centered at about the 40% point of the half-span, measuring from the root to the tip. With this assumption, the bending moment at the center of the wing, which must be carried by the spar is:

$$M = \frac{12 \text{ feet} \times 1000 \text{ pounds} \times 10G \times 0.4}{2}$$

$$= 24{,}000 \text{ foot-pounds}$$

(12 feet is the half-span. We divide by 2 because the other half of the gross weight is carried by the other half of the wing, multiply by 0.4 because the load is centered at the 40% point.)

The spar will resist this force-distance by creating an equal force-distance. With the tension in the lower cap and the compression in the upper cap acting over the distance between them. This is figured as the measurement between the **centroid** of the top cap and the **centroid** of the bottom cap. It is always a little smaller than the total wing depth. Because we don't know just what size the caps are, we cannot measure this distance. The usual practice is to guess at cap size and then go back and check to ensure that your guess wasn't too far off. We'll estimate that our 0.75 wing total depth will give us about 0.7, which will be accurate if the thickness of the caps comes out to be 0.025 feet, or about 0.3 inches.

Now, using the figure of 0.7 feet as the spar depth, we divide moment to be carried by the spar depth, and get the load in either spar cap, about 34,000 pounds in the top cap and 34,000 pounds in the bottom cap.

$$\frac{24{,}000}{0.7} = 34.000 \text{ pounds}$$

A spar cap that can carry 34,000 pounds in either compression or tension sounds like a lot of spar cap. Let's see just how much. Using E-glass fibers, and making a Burt Rutan-type layup with epoxy resin, we can expect around 48,000 psi in tension, and 40,000 psi in compression in the final part, if we keep the resin content low and keep the air out. Dividing 34,000 by 48,000, the cross-section area of the lower, or tension cap, is 0.71 square inches, and the upper cap is about 0.85 square inches. To calculate the spar cap depth, we would take the area of a cap, 0.71, and divide it by .3, or about 2.37 inches. If we use a nominal 2-3/8 inches, the small difference does not justify recalculation. The upper cap is then 2.84, or approximately 2-7/8 inches wide.

The above sizing calculation was for the maximum size of the cap, which is at the center of the wingspan. As the cap proceeds along the length of the semi-span toward the tip, the load gets smaller and the size of the cap can be reduced accordingly. The amount of the load at each point will depend upon the wing geometry, which affects the distribution of the air loads on the wing surface. As a first approximation, however, you can use a reduction that is proportional to the square of the location along the wing. For example, the load at the root is multiplied by .56 to obtain the approximate size needed at a point where 3/4 of the distance to the tip remains, as shown in Figure 9-2.

The reduction in material that is needed to carry the load can be made by either reducing the number of plies in the layup as the distance toward the tip progresses, or by making the width of the cap smaller as it goes outboard. If the reduced number of plies method is chosen, it is common practice to cover the stepped ends of the plies with an outer skin, or to make the full-length ply the outside layer. This will keep the thickness changes buried in resin inside the part. It is also helpful to put a caul plate (a caul plate is a plate which will bend to accept the curvature of the airfoil, but not enough to accept the dip into a low spot) over the finished layup and apply a little pressure, which forces a smooth surface on the outside surface of the part, and reduces subsequent finishing labor.

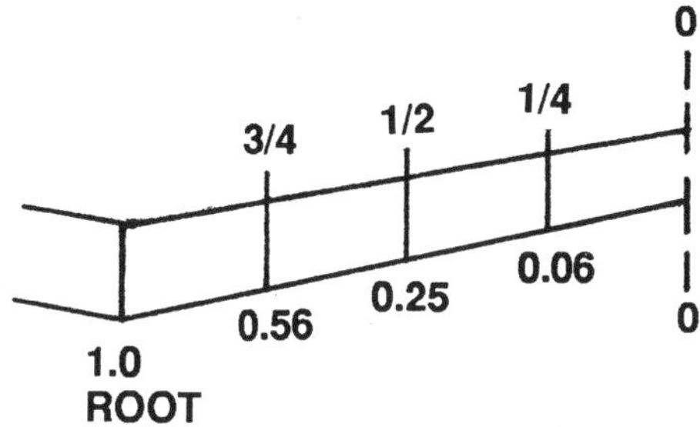

Figure 9-2.

NOTE: The analysis performed above is really only an approximation of the loads for which you must provide a structure. The best way currently to solve all of the many mathematical problems present in a complete wing design is to purchase or borrow one of the new wing design computer programs, and run a complete and accurate analysis. The simplest and easiest to use of those I have seen is that sold by Martin Hollmann, of Aircraft Designs, Inc., 5 Harris Court, Bldg 5, Monterey, CA 93940. Another alternative, for a really reliable structure, is to have Martin finish the design for you, after you have completed the preliminary design.

The example above was made of fibers of E-glass, and assumes a low resin content (about 50%) and an open layup without a vacuum bag. If we use one of the newer fibers to make the same spar cap, and employ a vacuum bag to achieve a lower resin content, it might save some weight and justify the higher expense and added work. If we choose S-glass instead of E-glass, the stress level that can be used for design is about 60,000 psi rather than 48,000, and a vacuum bag might get the strength up to 65 or 70,000. The unit weight of the laminate would be about the same, so the part weight reduction would be almost the same as the increase in strength, or about 25%.

If we use carbon fiber (and were able to achieve some of the advertised properties), the allowable stress level might be any where from 70,000 up to perhaps 160,000 psi, which would yield a needed cross-section of about .34. Because the laminate would have a density of 0.56 instead of 0.72, the weight reduction would be about 60%. Be cautious, however, in assuming that the allowable design stress used in the example above can be achieved in your parts. The actual layup technique and materials used have a **dominant** effect on the loads that the part will actually carry, and I know of one recent example in which a builder was sorely disappointed in the actual measured strength of an excellent carbon fiber spar layup.

This individual was an employee of a large aircraft manufacturer, (Boeing) and had access to their test lab. He designed spars in wet-layup carbon fiber material, and fabricated test coupons, which he tested for tension and compression, using the proper equipment and techniques. After two attempts at making suitable structures, his carbon fiber composite spar caps' best performance was barely over 60,000 psi! Upon reviewing the problem, he concluded that his layup procedure simply could not achieve the resin content and fiber straightness needed to hit 100,000 to 200,000 psi that he had read about, so he changed back to E-glass. He was able to locate a material that is used by makers of skis and compound bows. This material, a flat unidirectional laminate, is furnished in long, precured strips that have a sanded surface suitable for laminating or bonding without further treatment, and tested at well over 150,000 psi in both compression and tension. This material is sold as "Spar Tuf," by Gordon Plastics. This company has since added a high-strength glass fiber, RH Glass, from Vetrotec, as well as a carbon fiber version. Their address can be found in Appendix 1.

CAUTION: IF YOU CANNOT USE AN ESTABLISHED MATERIAL AND PROCESS COMBINATION THAT HAS BEEN TESTED TO ESTABLISH ALLOWABLE STRESS VALUES, DO MUCH MORE TESTING OF THE FINISHED PARTS YOURSELF. THE FIGURES ASSUMED IN THESE EXAMPLES WILL ALMOST ALWAYS BE DIFFERENT FOR YOUR PARTS, AND ONLY ACTUAL TESTS MADE BY YOU CAN FIX THAT NUMBER FOR YOUR SPECIFIC MATERIALS AND PRODUCTION PRACTICES.

HOLES AND JOINTS

One of the sore points in composites is the fact that most laminates, particularly the high-strength ones, are seriously degraded by drilling holes in them. The problem can show up as a substantial reduction in the part's strength. This reduction is always much more than that represented by the loss of area due to the hole, and is accompanied by poor edge-strength of rivets or bolts that are set into holes along the edge of a part, serious chafing of a bolt in a hole as it rotates or vibrates, exposing the fibers in the drilled hole to the subsequent ravages of moisture and time. Consequently, several design practices are often used by those knowledgeable in this area:

1. When rivets must be used in a laminate, they should be set into wet resin so that the inside cut fibers are re-sealed in resin, and moisture is prevented from entering the laminate along the fiber surfaces.

2. When more bearing strength is needed along the edge of a thin laminate, an aluminum strip of suitable thickness can be included in the layup. This piece must be properly surface-treated so as to ensure a permanent bond to the resin, or subsequent delamination may be expected with the passage of time. (See Chapter 11.)

3. If a heavy bolted attachment, such as a joint in a spar cap, must be provided, a large amount of excess strength must also be provided to keep stress levels below half of the levels in the undisturbed areas. Just make it twice as thick in the bolted area.

4. When the attachment of the above item is provided, it is also a good idea to provide a metal surface area for the bolts or nuts to scuff against, so that they will not later tear those structural fibers near the surface, either on installation or upon later service experience. In this case, the metal need not be properly prepared for bonding, as the bolt will continue to hold it in place when it does delaminate.

5. The best practice for heavy bolted joints, a practice now used extensively in the helicopter blade business, is to make the unidirectional fiber-spar structure into a multi-directional structure, as shown in Figure 9-3. This avoids the matter of cutting fibers, and allows the high allowable stress levels of the continuous fiber structure to be carried into the fitting areas and **around** the bolt holes, with no cut fibers at all. The best possible solution for a bolted joint for the spar-cap design would provide a structure such as this. Be careful, however, as this joint will carry very little compression load!

6. If many holes must be provided in a laminate, the material is laid up over a mold that has pins protruding from each place that a hole is desired. Fabrics are simply skewered onto the pins, and a protective pad is provided in the layup to protect the bag from the pins where bag molding is used. This method provides holes in the finished laminate with a minimal loss in strength, but be certain where you want the holes. They are **very** difficult to relocate after the part is cured.

ROVING WOUND UNDER TENSION

FIXED PINS

ROVING IS WOUND BACK AND FORTH ON A PIN WINDER OF SUITABLE LENGTH

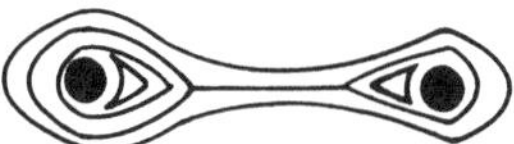

WOUND ASSEMBLY IS SQUEEZED INTO FINAL SHAPE AT ROOM TEMPERATURE STILL UNDER TENSION WITH PINS IN PLACE

COMPACTED ASSEMBLY IS TRIMMED TO TAPER FORM AND THEN CURED WITH HEAT AND PRESSURE

Figure 9-3

SPAR WEB

The previous example discussed a spar cap design, but did not describe a structure that holds the spar caps the proper distance apart. This function is provided by the spar web, and it is a rather straight forward problem. The spar web provides the compressive strength that keeps the caps from moving toward each other under load, and also carries the vertical and horizontal shear in the spar.

The shear to be carried comes from the same lift forces on the wing that cause the bending loads. The shear is different, however, and easier to calculate. At any point along the wing span, shear is the total lift that exists between that point and the wing tip.

For our airplane, the shear would be a maximum at the wing root, and would be exactly half the total load, since each half of the wing carries half the load. At our 10G and 1000 pound gross weight, the figure would be 5000 pounds. This load is picked up by the spar caps and ribs because they are connected to the wing surface, and the load is transmitted by them to the web. The web then carries the shear **uniformly** over its entire depth. If we make a sandwich web, the shear is carried by both faces of the sandwich, half in each face, or in proportion to their thickness if they are different. The arrangement is shown in Figure 9-4. The shear is being carried by the web and must be picked up at the fuselage from the web, not the caps.

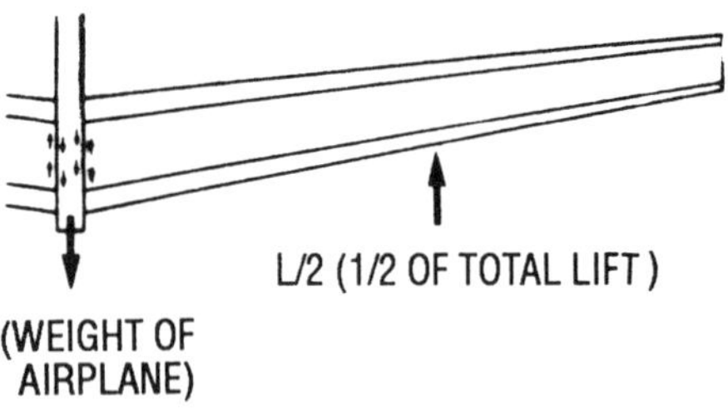

Figure 9-4.

To estimate the thickness of the spar web that is needed, we must know the allowable shear stress that the web material can safely carry. This value is often neglected in the data offered by the manufacturers, but for cross-plied composite materials, it is common to estimate its value as half the tensile strength. A spar web must be the full depth of the spar, but it does not carry much load. (Remember, we have about 34,000 pounds in **each** spar cap, but only 5000 pounds shear in the web.) It is common, therefore, to design the spar web from thin materials, and to make the materials into a sandwich so that they can be stabilized and kept from wrinkling at a low load.

If we again use E-glass as a spar web, and assume about 30,000 psi as the tensile strength of the fabric, the shear allowable would be about 15,000 psi. In order to carry 5000 pounds, we will need a total area of 5000 divided by 15,000 or 0.3 square inches of material.

In this case, we use the entire wing depth as the web depth since the caps also carry a little shear. The thickness of the web is:

$$\frac{0.3}{9(\text{spar depth}) \times 2\ (\text{two faces})} = .016$$

A double thickness of Rutan's BID would work well, or we might try a double thickness of 7781, which is a little closer and a little lighter. When designing with composites, we can make it a little too light, perform a static test and beef it up with just the required amount of material as a final step. In this case, we could add a single thickness of style 120, (which is .004 thick) to one facing in order to make a slight adjustment in strength. This adjustment to the web thickness is quite appropriate since we haven't yet determined the web thickness at some of the outboard points along the spar, where they will be thinner. Another good idea in designing spar webs is to orient the fabric at 45 degrees to the axis of the spar, which lends a slight improvement in strength because the shear loads are exactly aligned with the yarns, and a little lighter fabric can be used.

THE SPAR WEB CORE

The calculations needed to set the thickness and strength of the spar web's core are too complex and unreliable to justify using them, even if you are a good mathematician. The usual procedure is to simply pick a core type, density and thickness that works with these same facings and loads in other structures. We have determined that at the wing centerline, the point of maximum load, we need only a total sandwich facing thickness of .016 inches, which indicates that a core density the same as that used for the lightest sandwich facings would work well.

The chosen core thickness should be within a range of about 1/20 to 1/30 of the panel width, (the spar depth), or somewhere between 1/4 and a 1/2 inch thick. Remember that the thicker core will give sandwich stiffness that is six or seven times that of the thinner one, for bending, but not for shear. The only advantage of the thicker core is that the web seems to be "beefier." The other major consideration is that there is a lot of area to be dedicated to the spar web, and the thicker core will weigh more than the thin one.

SPAR WEB ATTACHMENT POINTS

One of the problems in working with composites that have a thickness of .016 inches is attaching things to them. In our example, we must provide for a sturdy attachment to the fuselage, and perhaps also to the landing gear or some other substantial fitting. The method is always the same. Be sure that the loads are being carried in the plane or in the direction in which the structure is oriented. If the direction of the load is wrong, or if you cannot be sure of the direction, simply make a surface-applied fitting that picks up the facings, the core and the caps in a single mass, and you can make almost anything work.

If a hole must be cut in the spar web, reinforce the area around the hole with a little more than that which was cut out, including an extra piece of facing that extends through the hole and onto the skin on the other side. If the web must have a fitting or bracket set on its surface, add another piece of core and another skin so that the surface of the web is built up, away from the original surface. If you use a heavier core, such as spruce, you may be able to sink screws into the pad. If bolts or screws must go through the web, cut a piece of the core and one skin away as shown in Figure 9-5, inserting a heavy enough core so that nut tightening loads will not crush the core.

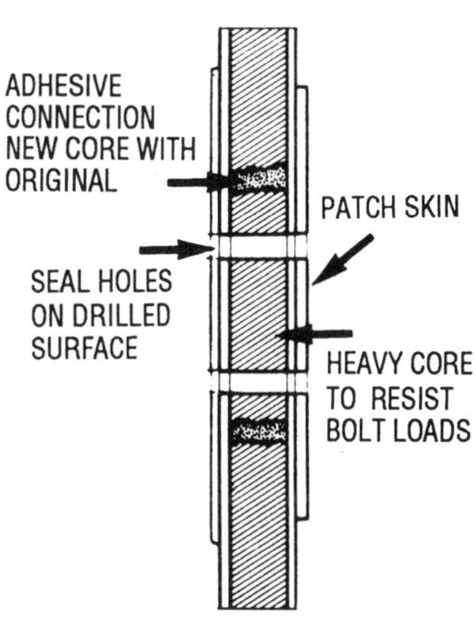

Figure 9-5.

As an alternative, a metal bushing can be glued into the hole, which takes the bolt tightening loads. You may still need the heavy core, however, as such fasteners usually have a larger bearing load requirement than the light cores and thin facings will carry without failure. If the bearing loads are very high, you can also enlarge the diameter of the flange on the bushing, so that it transfers the bearing load from the bolt into the tensile or compressive load in the facing.

ATTACHING THE WEB TO THE CAPS

If the spar web is light, such as the one chosen for this example, it can be attached to the spar caps by using a glued joint with a corner reinforcing tape of the same or heavier fabric as the web facings. If, after assembly, access to both sides is difficult, a tape from the far side can simply be brought across to the near side, as shown in Figure 9-6.

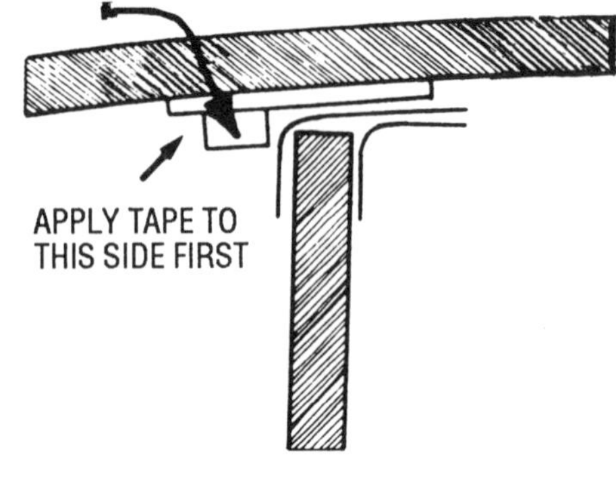

Figure 9-6.

If your calculations indicate that only one or two thousandths of an inch of spar-web facing are needed, the tape can be eliminated entirely, and the bond of the spar web's core, even if it is only 1/4 to 1/2 inch thick, will be sufficient to transmit the load from the web facings into the cap.

THE WING SKIN

One of the best ways to design a wing skin for our example airplane is to make it a sandwich. The outer facing is made much heavier than the inner, as the total of the two together are usually much more than is needed.

When even a light facing is put on the other side of a slice of core, however, the resultant sandwich is usually stiff enough so that most of the ribs and other internal structures are no longer needed. The skin can also be used to take the wing torsional loads, so that no internal drag bracing is needed for this loading condition. To design a skin that is strong enough to take these loads involves merely providing stabilization for the minimum thickness that will be chosen for puncture resistance. Because the area of the wing skin is so large, the loads themselves are usually an order of magnitude lower than even the spar shear loads, and become important only when the speed or performance of the airplane rises to rather high levels. Loads **must** be calculated later, but only as a final check to verify that the chosen skin thicknesses are adequate.

An adequate thickness for both a "solid feel" and for puncture resistance depends upon your personal preference. Using glass fabric and resin, the threshold at which the skin can be punctured by a sharp pencil is about 0.015 inches thick. A single ply of Rutan's BID is about this thick, while a single ply of the Glasair`s 7781 is only about .009. I suspect that the two-ply outer skin on the Glasair was chosen specifically to achieve a solid feel rather than as a structural requirement. The inner skin can be much lighter than the outer skin, provided that the core is thick enough to arrive at the necessary overall stiffness.

The necessary balance between inner skin thickness and core thickness will depend upon the minimum core density you use, as even a 4 pcf core will add considerable weight when the thickness reaches an inch or so. Also, the minimum available thickness of glass fabric, around 0.001 inch, will need a disproportionate amount of resin to bond it to the core and stabilize the fibers. As a practical matter, the lightest weight glass or Kevlar fabric that you can usually use is a 120 style at .004 inch, or perhaps a 112 style fabric at .003 inch. The lighter weaves, some of which are .001 inch and .07 ounces per square yard are almost the same weight as the 112 or 120 after the needed resin is added. As a realistic compromise, your skins will usually need at least a 1/4 inch thick core and a 3 mil fabric on the inside surface, just to keep the handling damage at a tolerable level.

An outer skin of a single ply of 7781 will work, but you will have to put up with an occasional friend, (former friend?) putting a pencil through it. If your project can stand the extra cost of a Kevlar outer skin at .009 inch thick, it will save a little weight and be substantially more puncture-resistant. A weave comparable to a glass 7781, however, is quite expensive. One of the coarser weaves, style 285 or 500, yields the same performance, and costs about half as much, provided that you can live with the coarser finish and the subsequent added finishing labor.

FITTING DESIGN

The two ever-present problems in designing fittings in a composite structure are the fitting itself and ensuring that the structure that the fitting ties into is strong enough to carry the loads being dumped into it.

To illustrate how these problems are handled, let's design a seat belt attach point for a two-place side-by-side configuration. For our example we will assume that there is no main wing spar handy to take the loads, and they must be carried into the fuselage shell, which will then transfer them to the wings at some remote point. (This would be the case in a Quickie II or a Dragonfly, as well as with the typical biplane.)

The steps to take in this design procedure are as follows:

1. Find the location of the load.
2. Find the direction, or several directions in which the load will be applied.
3. Determine, or assume, the exact number of pounds that make up the load, and choose an appropriate safety factor.
4. Decide what sort of fitting you wish to attach the load to, i.e., a bolt, a stud, a hole for insertion of a bolt, etc.
5. Decide on the allowable stress level that the material you plan to use will develop.
6. Add layers of material to arrive at the total amount that will handle the load.
7. Look over your complete layout. Would a different arrangement do a better job?
8. Redesign it again, and then again, until it looks right.

9. Sketch up the fitting design, using enough plies of material so that the structure will carry the planned load, and with the orientation of the plies such that the strength **in all of the load directions** will exceed the maximum load expected in those directions.
10. Examine the attachment of your newly designed fitting to the back-up structure, making sure that the loads will be carried through into this back-up structure.
11. Analyze the back-up structure to determine that it will carry **both** the newly introduced fitting loads **and** the loads that the back-up structure was already built to carry.
12. Design a beef-up for the back-up structure if you are introducing more load than the total it will safely carry.
13. Follow the path that load will take to determine that the structure that the back-up structure feeds into is not overloaded, and design a beef-up for this structure where needed.
14. Where the loads to be carried by the fitting you are designing are substantial, repeat step 13 to follow the load path all the way to where the structure is clearly capable of carrying this added load entirely within the original design strength, or to where the load is either taken out into the air, or into the ground. (The exception, of course, is where the fitting is to carry sudden-acceleration loads, towing or pushing or lifting loads and the like, which only result in a change of the speed of the structure, rather than an equilibrium state of flight or resting on the ground.)
15. If anything doesn't look right on **any** of the above steps, back up and do it over a different way. (This is so standard in most engineering design procedures that it even has a name—**iteration!**)

Now that we have some steps to follow, let's make some assumptions about the structure in which our fitting will be used. For starters, Figure 9-7, on the next page, shows a section through the fuselage just forward of the back of the seat bottom. (For purposes of illustration, all of the usual consoles and side mounts have been eliminated from the sketch.) Again, for design illustration purposes, we will assume that the fitting is to carry a **25-G load**, the new and higher figure that the FAA is discussing. Let's also assume that each attach point will carry half of the standard 170 pound person, and that if we elect to add a fitting to attach a shoulder harness or chest cross-strap, that it will also be sized to carry half of the 170 pound weight.

There are several possibilities for location of the fitting that will go between the seats and carry half of the belt for each seat. These might be at such locations as:

- on the floor,
- on the seat bottom, if it is a bench seat,
- on a console, if there is one located between the seats,or,
- on the seat-back bulkhead.

SEC B-B SECTION THROUGH SEATS

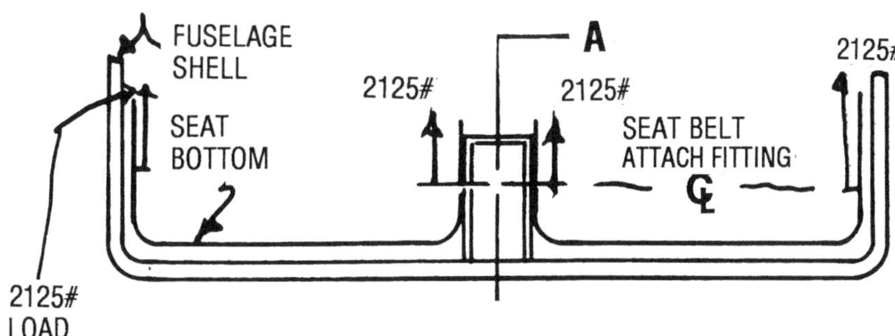

Figure 9-7.

CASE ONE: FLOOR LOCATION

Let's take the above possibilities in order. In this situation, we have a large flat panel, the fuselage bottom, which must carry a load of about 4250 pounds (170# x 25 G), normal, or nearly normal, to the plane of the panel.

In a flat panel, the portion of the loads normal to the surface must be carried by the core acting in shear, or by some other shear-capable member. It is common to replace the foam core in the area of the fitting with a strongback, or other solid insert, which can both distribute the load over a larger area, and carry the large local load of the attach-point. In this case, the foam core will only carry about 40 psi in shear, so a one inch thick panel would require 4250/40, or more than 106 inches of length of attach area for the foam to carry the total load. Both sides of the core next to the insert can carry this load, so the 106 inch length translates to 53 inches of fitting length. This is rather ridiculous for the size of the floor in this area, and it would also lead to other secondary loads and floor deflections that are really out of context for the sort of structure we are looking at. Figure 9-8 illustrates the problem.

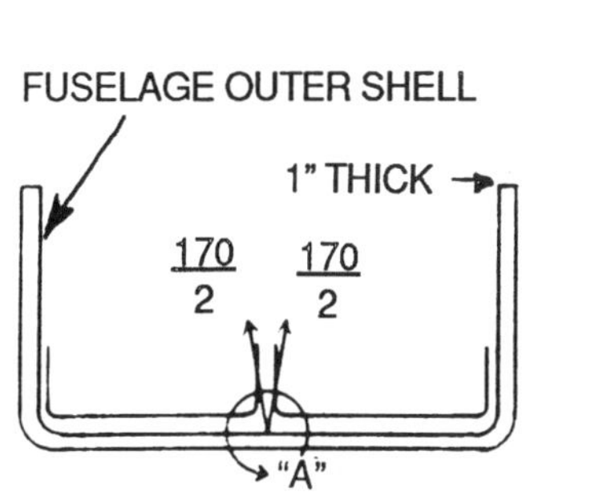

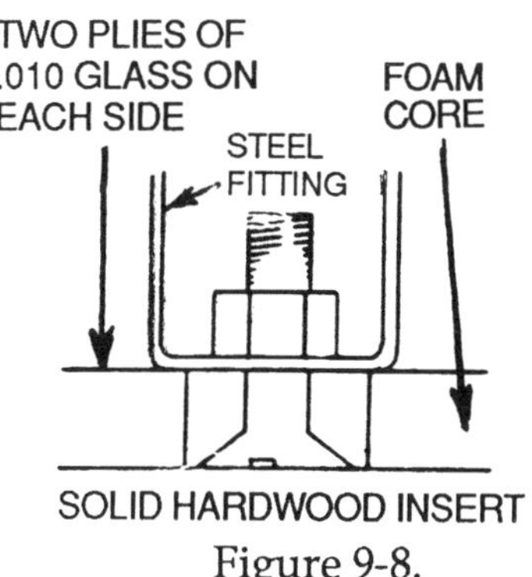

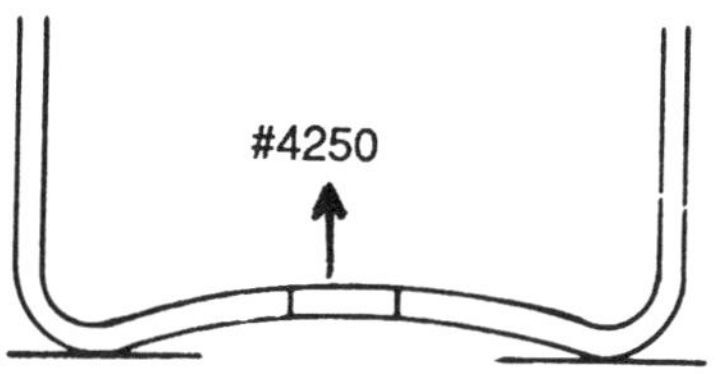

Figure 9-8.

CASE TWO: A SEAT-BOTTOM LOCATION

The seat's bottom is usually about the same strength, or lower strength structure, as that of the fuselage shell. For this reason the seat bottom will probably have the same trouble as the fuselage bottom in carrying the large load we have chosen for our attach point. The exception would be if the location were close enough to the seat back, so that the seat back structure could be used to carry the major portion of the load from the fitting.

Sometimes there is also a console between the seats, and because this will, or can be made to, attach to both the floor and the seat-back bulkhead, it is one of the best sources of potential structural strength.

CASE THREE: THE CONSOLE

The console location offers the capability of splitting the loads into two parts, each one just half of the 4250 pound figure, or 2125 pounds each. Each load is to be carried by a side-wall of the console. It also has the potential of carrying the loads to both the floor of the fuselage and the seat-back bulkhead at the same time, so that the structure has a closer conformance (not much beefier) to what it would have in its original form. The layout might look something like Figure 9-9.

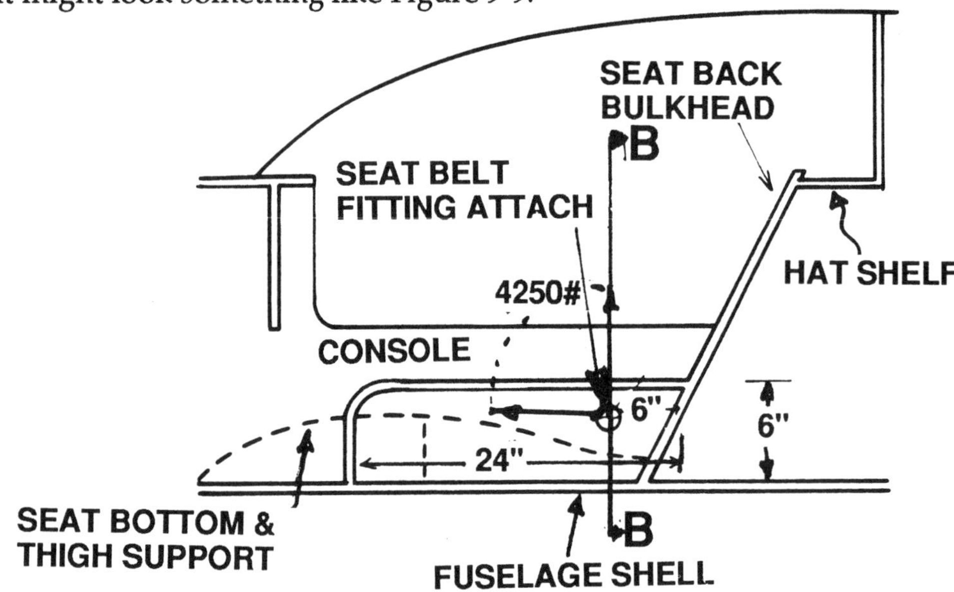

Figure 9-9.

Looking at Figure 9-9, we can assume that the load is parallel, or nearly parallel, to the sidewall of the console, so it follows that the entire load is carried by the facings of the console. These facings have been stabilized by making them into a sandwich structure with a foam core, just like the rest of the composite airplane structure with which we are working (a Dragonfly or a Quickie II). We must also assume some properties for the materials in the structure we are to build, as well as for the structure already there that we will use to mount our new fittings on. The values can be found through testing, but that is impractical for the already existing parts, so let's just assume that they are typical homebuilder structures with the following values usable for design:

Core: Urethane Foam, 4 to 5 pound
Compressive strength = 60 psi
Shear strength = 40 psi

Facings: 7725 (RAF BID) glass fabric, .010 inches per ply, with about the following values usable for stress purposes:

Compressive Strength = 25,000 psi,
or 250 pounds per inch per ply

Tensile Strength = 30,000 psi
or 300 pounds per inch per ply

Shear Strength = 10,000 psi
or 100 pounds per inch per ply

Bearing Strength = 50,000 psi,
or 500 pounds per inch per ply

Resin System: Derakane 411-45, with about the following usable values:

Interlaminar Shear Strength = 5000 psi
Secondary Bond Shear Strength = 1000 psi
Secondary Bond Tension Strength = 100 psi
(just like rivets, we try not to design for **any** load in the tension direction)

All of the above assumes that values should be reasonably close to what you could expect on your project, but may or may not be the actual values you will get when you load your finished fitting to failure and actually measure the strength. The strength of the secondary bonds where you make a new laminate that attaches to the old structure, which may be several years old, is a particular problem. The values shown would be reasonable if the original structure were made of the same resin system and did not contain any wax (commonly added to this sort of resin to make the outside surface cure up hard and non-tacky), but would possibly be a little lower if the old structure were one of the epoxy systems.

If the original part had been made of a polyester resin system, all bets are off, and you will be lucky to have it develop any usable strength at all! The standard approach is to laminate a small test-patch to the old structure, and after it has fully cured, pull it off in a peeling mode, looking for both a tough bond that is difficult to peel, and a lot of glass coming off with it. In other words, the secondary bond should about be on a par with the original structural laminate, if you are to trust the structure. Be careful, however, as an excellent secondary bond means that you may destroy the original structure when you make the peel-off test.

Looking at Figure 9-9 again, we see that the console is sort of a box, with each side a vertical panel having a single ply of BID on each face. If we assume that attachment is to be through a long 1/4 inch AN bolt, the load of 2125 pounds will be picked up by both facings, so we can assume they will each carry half of the total. The size of the console will have a direct bearing upon whether it can carry these loads, as we see in a moment. First, however, we must design a pad, or strong-point into which the bolt can transfer its load without destroying the immediately surrounding structure. Figure 9-10 shows how this may be done.

The flanged bushing is provided so that the bolt may be inserted and removed many times without scuffing up the hole and getting sloppy. Also this permits us to tighten the bolt without crushing the core in the sandwich of which the console is made. For a simpler installation use a single, long bolt to pick up both belt fittings, and **do not** tighten the nut any more than "hand-tight," unless you have provided a sleeve between the two sidewalls to take the compression.

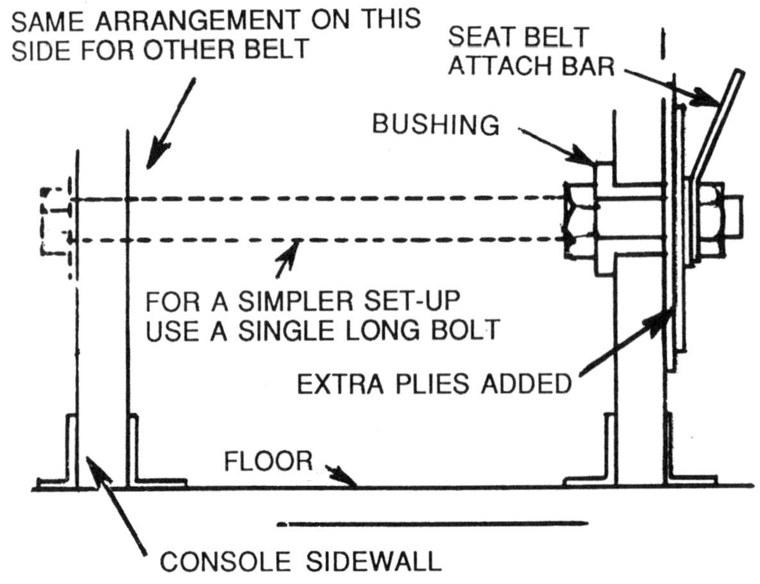

Figure 9-10.

Since the location of the fitting is only a few inches forward of the seat-back bulkhead, most of the load in the fitting will be transferred to this bulkhead by way of a path that goes through the console between the fitting and the seat back. The number of plies needed around the hole is determined by first calculating the amount of material needed to resist the bearing load of the insert on the edge of the hole in the plies through which the insert goes. For the insert shown in Figure 9-10, the hole will be 1/2 inch in diameter, so the area needed will be:

$$\text{Diameter} \times \text{Thickness} = \text{Area, or } T = A/D$$

Also the area needed is calculated by dividing the load to be carried by the allowable stress: A = 2125/25,000, or A = .084 square inches. If we substitute these values, then the thickness of the plies is:

$$T = A / D$$

$$= .084 / 0.5$$

$$= .168 \text{ inches, or 17 plies}$$

Note that no safety factor has been used in the calculations so far. At this point you should decide whether the added factor is needed, or whether the 25-G load rating itself constitutes a sufficient safety factor. If you really want to make good a 25-G load in a static test, it would be a good idea to include a safety factor in all of the calculations. The normal factor used for metal fittings of this sort is 1.15, but for hand-laid-up composites, a factor of 1.5 is a minimum, and a factor of 2.0 is often used where safety-of-flight items are concerned. Because we want to make good the target 25-G load, let's choose a factor of 1.5. It doesn't seem worthwhile to go all the way to 2.0 as a factor. (The seat belt starts to cut you in half at about 25 G's, anyway.)

This means that our 17-ply pad should be 1.5 times as strong, or be a total of 25 plies, 12 on each face of the console sandwich. To reduce this to a lower number of plies, simply use a larger diameter bushing. If you use a one inch diameter bushing instead of 1/2 inch, the ply count will then be about 12 instead of 25.

We should also figure out where this 2125 pound load that we introduced into the side-wall of the console will go. Looking at the side view of the set-up shown in Figure 9-9, we see that the console is 24 inches long and 6 inches high, and that the seat belt attaches about 6 inches forward of the seat-back bulkhead. For purposes of analysis, we will consider that the console is a box beam 24 inches long, with the bottom edge continuously attached to the floor, and the aft end continuously attached to the seat-back bulkhead. As a console, however, it probably wasn't designed to take the load of 2125 pounds on each side that we are now asking it to carry. We will plan, therefore, to add enough material to ensure that it does its new job.

Looking at the direction of the load in Figure 9-9, we note that a small component of the load will act parallel to the floor, while a much larger component will act parallel to the seat-back bulkhead. To arrive at the exact number of pounds, lay out a drawing to scale, and measure the loads. By eye-balling them, it appears that the horizontal part will be about 400 pounds and the portion parallel to the seat back will be about 2000 pounds. The panels will not pick up much load in any direction other than their own plane because they are several orders of magnitude stiffer in their own planes than they are when bending in a direction normal to that plane. The side of the console will carry load in its own plane, so it just cannot move enough to load up another panel that it is joined to, unless that panel is also stiff.

It is always a good idea to guess at the relative stiffness of the various panels that you are connecting to carry a load, as the **load will always follow the stiffest path.** The more flexible structure simply moves over without picking up much load, and makes the stiffer structure next to it do the job. If you have not provided enough strength in this stiffer part, it will simply fail without alerting you in advance.

Before we worry about transferring the load to an adjoining structure, however, let's make sure that the structure with the fitting is strong enough to transfer this load over there. In our case, we have provided a total of 12 plies on each side of the console in order to pick up the load in the fitting itself. Now we must look at the rest of the console sidewall structure to see how far along the length, and above and below the bolt, these extra plies must extend.

The load will be carried in shear in the panel facing, and as Figure 9-9 shows, the panel is 6 inches deep, which means that each of these 6 inches must carry 1/6 of the total, or 2125 divided by 6, or 354 pounds. When we decided on the figures that we would use for design allowables, the number we picked for shear strength of the facings was 10,000 psi. Therefore, each inch of facing must have a total thickness of 354, divided by 10,000, or .035 inches, to make this work. Also, we use a safety factor of 1.5 as discussed above, so the thickness needed is 1.5, multiplied by .035 inches, or .053 inches total. Our plies of facing only come in multiples of .010 (for the fabric style we are using), thus, the side wall away from the immediate area of the bolt hole must have a total of **six** plies of fabric in the facings. It already has one facing on each side, so we will need to add four more on each side.

But hold on! The load of 2125 pounds will be carried partly by the seat-back bulkhead, and partly by the forward floor of the fuselage. If we look at Figure 9-9 and treat the console as if it were a whiffle tree, (see Chapter 11 for ratios, levers and whiffle trees), we can calculate that the seat back end will carry a load equal to the primary load, 2125 pounds multiplied by the lever ratio, 18 divided by 24, which is 1594 pounds. This same ratio, 0.75, can be used to reduce the total number of plies or pounds to be carried. Since the ply count contains many rounding-off errors, let's use the calculated thickness to modify the total; this is .053 multiplied by .75, or .040 inches exactly. This is exactly four plies, which is the **total** number of plies needed in the six inch deep area extending from the fitting back to the seat-back bulkhead. Thus, we need add only two more (and they can both be on the **outer** facing if it is more convenient).

To calculate the facing requirements forward of the fitting, we should first look at the relative stiffness of the panels. In reacting to the straight-up direction of the load, the console sidewall is **stiff** when compared to the flat floor that will have to pick up the load, so most of the load will probably get all the way to the front end before it is taken out by the reaction of the floor. Also, this reaction must be provided by the facing of the floor, unless we insert a fitting into the floor for the entire length of the console—a messy job.

Instead, let's assume that the floor will not carry any of the load, except at the forward end of the console, and provide a special fitting there to pick it up. The console sidewalls will need to be the same strength all the way to the front end. This will be calculated by the same ratio calculation as used for the aft end, except that the ratio is 6 divided by 24, or .25. Note that this number, plus the ratio calculated for the aft load, makes a total of 1.0, which confirms our arithmetic. Using the same method as used above for the aft portion of the console, the thickness needed is .053, multiplied by .25, which equals .0132 inches, or two plies. Since we already have two plies as the facings of the console, no extra plies are needed from the fitting forward!!!

The total that we must take out at the forward end is still of some consequence. It is calculated by multiplying the original 2125 times two (since we have a fitting on each side of the console), and reducing this total by the same .25 factor used above. This number is 1063 pounds, which is much more than can be picked up by the fuselage floor facing in flatwise tension. We will, therefore, need to either install a fairly stout fitting in the floor so that it can pick up this load without failing, or we can carry the console further forward and join it to the instrument panel. The instrument panel bulkhead is rather like the seat-back bulkhead in that it is on edge instead of flat like the floor; it doesn't take much of a panel to easily pick up rather large loads and without much change in the structure that you would use in the absence of such loads.

Looking at Figure 9-9 again, it would appear that this would make the console about 6 inches longer, and the instrument panel bulkhead would have to extend about 8 inches farther down than is shown in the sketch. This design change would also affect the amount of load carried by the seat-back bulkhead and the forward end of the console, changing these to 0.8 for the aft load, and 0.2 for the forward load.

If we move the location of the fitting 2 inches closer to the seat-back bulkhead, the factors would be 0.87 and 0.14! (The total is greater than 1.0 because both are rounded off to a larger figure.) This will then require more plies than we had initially calculated, so that the larger load can be carried from the fitting back to the rear bulkhead. The weight would be about the same, however, because the extra plies don't have to go so far. The ply count would now be .87 times .053 inches, or .046 inches. This translates to 5 plies, because of our .010 inches of thickness per ply. (Remember that the console side wall **already** has two plies on it.)

Our seat-back bulkhead was intended to be just a fuselage former and a backrest, so we will assume that it has only a single ply of our .010 inch thick BID as a facing on each side. This is the structure that we are asking now to pick up approximately 2000 pounds which we are transferring from the console side wall, remembering that there is that same 2000 pounds in the other side of the console also, for a total of 4000 pounds to be transferred.

The single-ply facing is .010 thick, and we multiply this figure by the allowable stress and divide by the safety factor, which gives us the number of pounds that will be carried by each inch of the facing of the bulkhead—67 pounds. That means that the bulkhead will have to be 4000 divided by 67, or **sixty inches high.** This is more than the depth of the fuselage, so we will have to carry the load in more than one ply of thickness. One extra ply would reduce this depth to 30 inches, but the bulkhead is only about 24 inches high, if we are to believe Figure 9-9.

Again, however, half the load will go to one side of the fuselage shell, and the other half to the other side. We must carry the 5 plies on each side of the console on to the face of the bulkhead, making them larger than the 6 inch depth of the console, and reducing the number of plies as they get larger. At a 12 inch depth (measured vertically on the seat-back bulkhead), we would be at 2.5 plies, and at 18 inches, the existing 2 plies of the bulkhead structure would take the load by themselves.

THE SIDE-WALL FITTING

The side fitting is a little easier than the center one, since only half of the weight of one of the passengers will be taken by each fitting. The load will be the standard 170 pound FAR person, multiplied by 25 and divided by 2, which is 2125 pounds. The direction of this load might be anywhere from straight up to straight forward just as it was in the center fitting.

In the case of the center fitting, however, the two loads tend to balance out to zero for any sideways component, assuming that both seats are occupied, so that the nearby fuselage structure will not see any side load component at the same time that the entire load imposed by two people is applied. In the case of the side-wall fitting, however, if the load is applied, it will always have some component going sideways toward the opposite side of the fuselage.

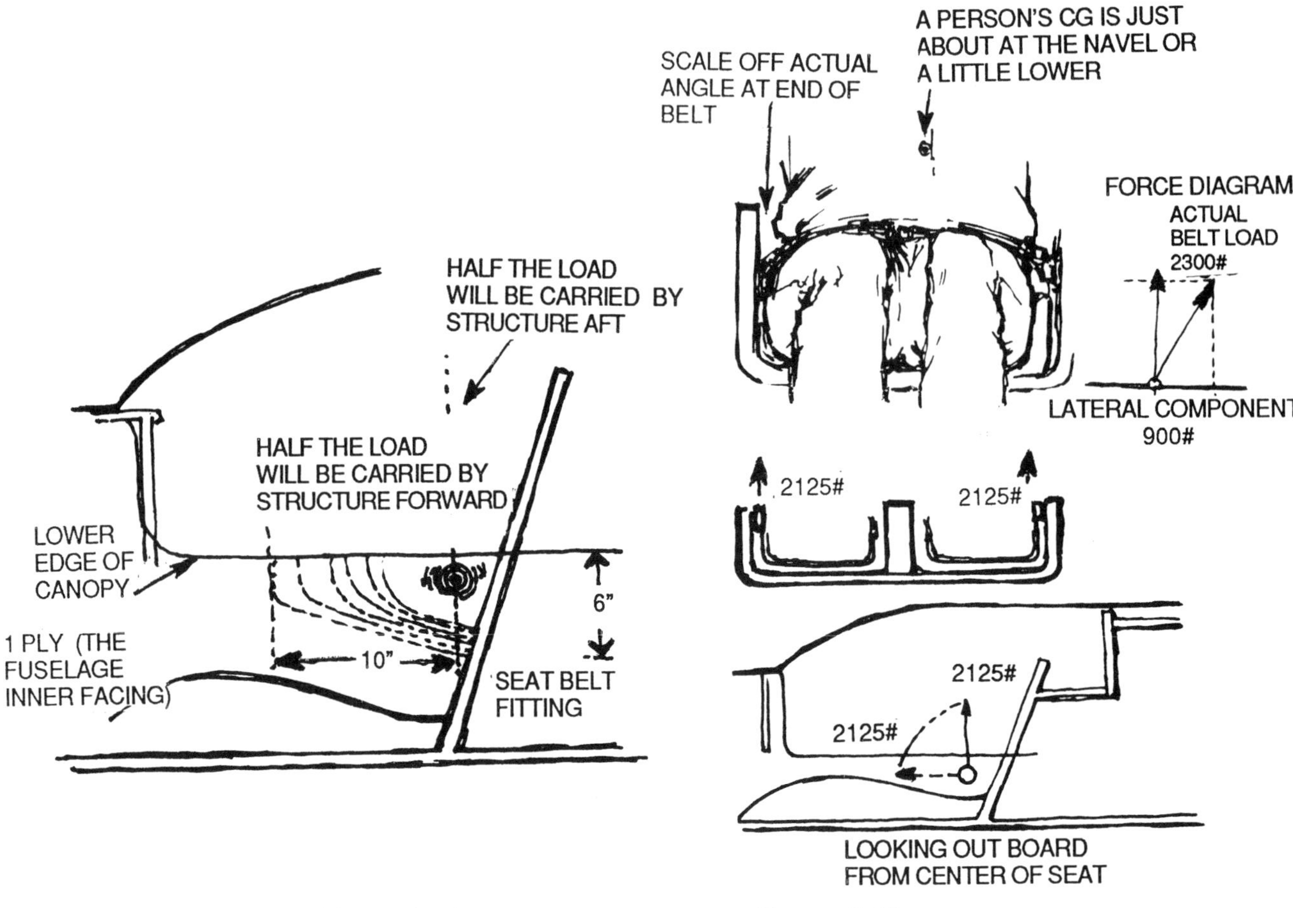

Figure 9-11. Figure 9-12.

If we look at Figure 9-11, it looks as if the angle would give an inboard-acting component about half of the primary load, or about 1100 pounds. Remember that you can find this value quite accurately by laying out a drawing to exact scale, including the exact size of your seated posterior, and just scale off the load values for the unknown loads in strange directions.

Figure 9-12 shows how you might actually do it, and if my drawing fits your case, the part of the load acting straight inboard will be about 900 pounds. (This is typical of the difference between the value you guess at and the true value, whether scaled or calculated.) Note that the true load on the belt is actually higher than the 2125 pound up-load.

Analyzing for the lateral component of the load first, we note that the 900 pounds must be carried by the side wall of the fuselage, but that the plane of this part of the fuselage shell is at right angles to this component of the load, which means that we will have the same problem that we had in the center fitting when we tried to attach the fitting directly to the floor. The load in this case is not quite as high, 900 pounds instead of 4250

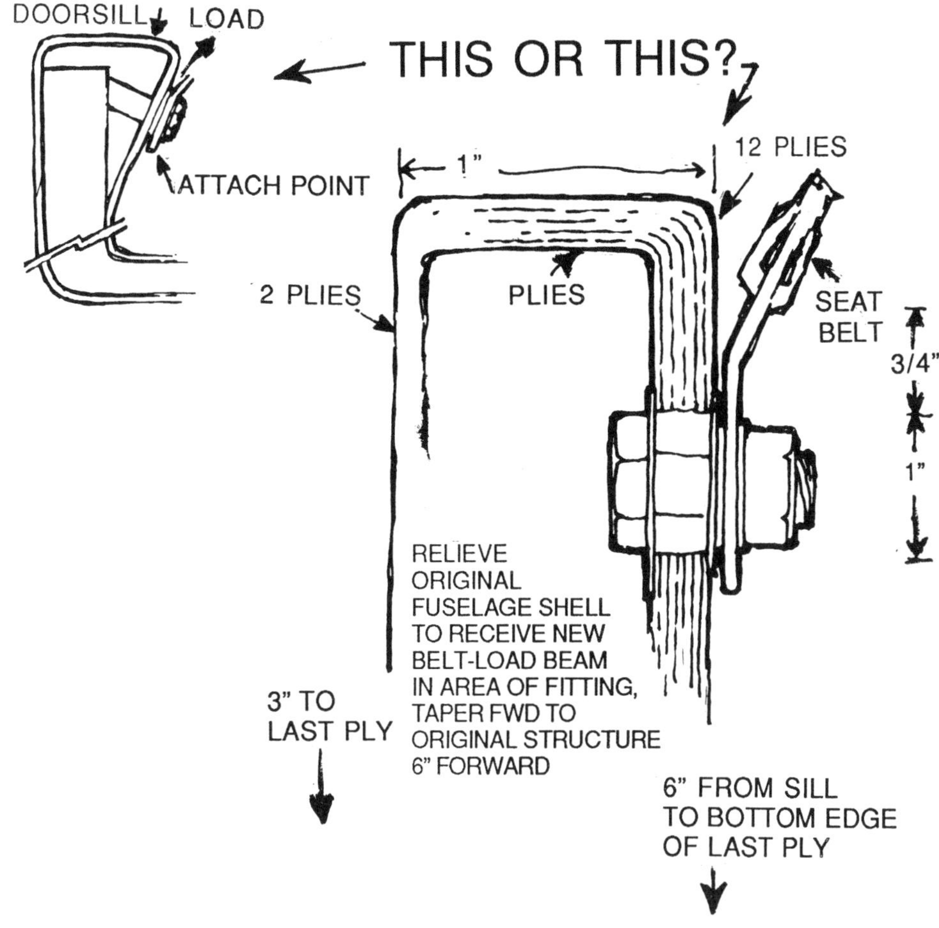

Figure 9-13.

pounds, but is still a rather large load for a sandwich panel to carry normal to the facings. Using the same 40 psi shear allowable for the core, with a 1 inch thickness, the fitting we would insert into core would need to have a total edge length 900/40, or $22^1/_2$ inches.

Assuming that the core on both sides of an inserted rib would pick up the load, the rib would then work out to be about 11 inches long. Since this is not so preposterous as the length we calculated for the insert in the floor, we might seriously consider it. However, it would be easier to locate the fitting higher, where the lower line of the canopy cut-out forms the top edge of the fuselage shell in this area. This door sill could then be strengthened to take the inboard component as a beam in bending, while the primary load, up or forward, could be taken by the facings on the fuselage shell.

The door sill beam would best be in the shape of a letter "C" with the legs pointing down, and the full fore-and-aft length of the entrance area. Since the sill area must have an extra layer or two of fabric just to withstand normal door-sill abuse, this design might turn out to be practical, after all!! If we also make the flat back of the C-channel wider than the 1 inch thickness of the canopy cut-out, we can make the inside leg of the channel go back to the original inside skin of the fuselage in order to provide the proper angle at which the seat belt should leave the attach area toward the side of your lap. Figure 9-13, on the next page, shows how it might work.

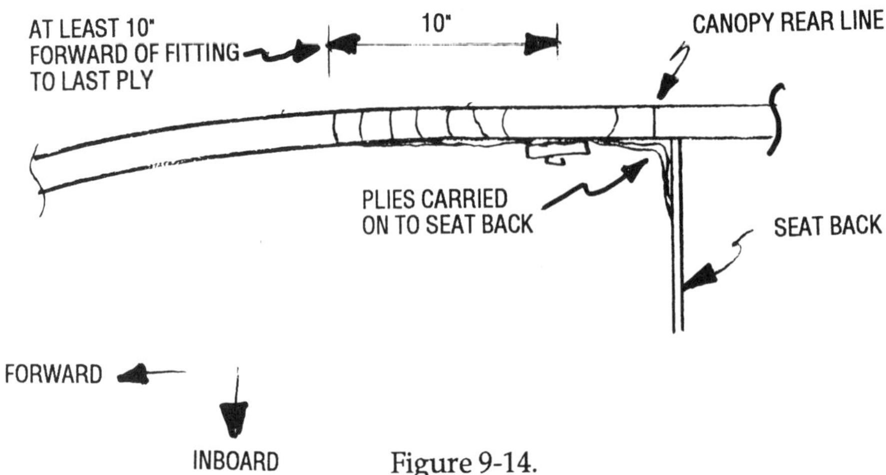

Figure 9-14.

One possibility is shown in Figure 9-13, in which the dimensions are already fixed, i.e., the 1 inch thickness of the fuselage shell, the back of the "C" channel that will form the exposed surface of the door sill beam, and the angle of the inner face, where the actual fitting will be located. This full scale layout will allow you to obtain the rest of the cross-section dimensions that are not already fixed. Be sure to include such details as the flanged insert you will use, the belt-end fitting that will attach to the structure you are designing, and so forth. For example, if the metal end-fitting on the belt remains above the level of the top of your door sill when the belt is unfastened and the seat is unoccupied, the sill will be uncomfortable for anyone sitting on it while getting in or out of the airplane.

Figuring out the thickness or number of plies needed will follow the same procedure we used in designing the center console attach points. Assuming that we will bury a bolt in the structure so that a threaded end sticks out for attaching the seat belt, we will need the same size flanged fitting in the glass for providing enough bearing strength. The same 1 inch diameter fitting will work, as the load on the belt is the same as before. Also, this diameter will require the ply count to be only 12 at the fitting area. Whether we use a threaded fitting with a stud sticking out for the attachment of the belt, or a through fitting with a bolt head on the inside, the fitting will need a flange on the back side in order to resist being pulled through the glass layup by the 900 pound lateral load.

To calculate the size of this flange, use the shear allowable for the metal you will use to fabricate the fitting, and divide the 900 pound load by this figure. If we use 2024T-3 aluminum bar stock for the part, we can expect the shear strength to be about 40,000 psi, so the flange will need a cross-sectional area of 900/40,000, or 0.022 square inches. This flange, or ear, will be a diameter larger than the 1 inch basic diameter of the fitting, so the area under shear load will be the circumference of the fitting, multiplied by the thickness of the flange. In order to keep this flange from being fragile, it should be at least .020 thick, which would provide a shear area of 1 inch x 3.14 x .020 inches, or 0.0628 square inches.

This means that our .020 flange will have about three times as much material as it needs in order to carry the pullout shear load at the flange. With this amount of excess strength, we won't bother with the usual 1.15 fitting factor.

The bearing strength of the glass laminate resisting the 900 pound load that the flange imposes on it must also be considered. Earlier in this chapter, you will recall that we assumed that the laminate would have a bearing strength of 50,000 psi, so we can calculate the height of the flange quite easily. The area is equal to the height multiplied by the length, 3.1416 inches (the 1 inch diameter times pi), and we simply use old faithful formula, $S = P/A$.

S is the stress, or 50,000 psi, P is the 900 pounds we are to carry, and A is the flange area noted above, or 3.14 times the height. Since we are not interested in any particular height, let's make it $^1/_{16}$ inch and see if it is adequate. 900 divided by 50,000 is equal to 0.018 square inches, while a $^1/_{16}$ inch high flange is 3.14 times .062, or 0.196 square inches in area, about 10 times as much as we need. Almost any little bump on the back side of the fitting would probably be adequate to resist the 900 pound pullout load!

Now that we have sized the fitting that we will bury in the laminated part when it is finished, we can calculate the thickness of the beam's flanges, the thickness of the web, and the length of the various plies, all the way to the full length of the door sill. Figure 9-14 shows some of these critical dimensions.

Note that the inside surface of the channel is shown as a vertical leg rather than the inward sloping leg mentioned earlier, and shown in Figure 9-13. The reason for changing it back is that the 1 inch fitting diameter makes the top of the leg jut too far into the seating area for comfort, and the pullout load we just calculated above is so trivial that it is not worth making our seating width smaller. Also the extra thickness of the many plies in the new structure are carried inside the 1 inch dimension of the fuselage shell, so the installation does not look so bulky as it would with all those plies scabbed on the outside.

Remember that we have 12 plies in the area of the aluminum insert; they all will pick up part of the other loads, too. If we drop off one ply every half inch, starting an inch away from the insert, we can use the same calculations we used on the center console design. The main difference is that we are picking up the load on the fuselage shell directly. Figure 9-13 shows what the layout will look like for the inside surface. As we go forward on the inside skin, the load will gradually transfer to the outside skin also, so that within 10 or 15 inches forward, we can count on the outside facing of the fuselage sandwich to be carrying about the same load as the inside facing. This gives us two skins, or a shear capability of 200 pounds shear per inch of length. Anything more than 11 inches can then be expected to pick up all of the 2125 pounds in the fitting; even 5 1/2 inches would pick up the half that the portion of the structure forward of the fitting will probably be carrying.

Yet, the outside facing requires some distance to pick up its load through shear transfer of the core, so allow about 8 to 10 inches total distance from the fitting centerline to the forward edge of the last doubler ply.

The aft portion of the structure is just like that forward of the fitting centerline, except that the plies wrap around onto the forward facing of the seat-back bulkhead, just as in the case of the center console design.

The horizontal surface that is the door sill in the area of the new beam is also the shear web of the beam which will resist the inward-acting component of the seat belt load. This 900 pound force must be carried by enough plies so that the 1 inch width of the beam can pick it all up and transfer half to the forward part of the fuselage shell, and the other half to the aft structure. The load will get into this shear web by the flange of the fitting bearing on the 1 inch diameter hole in the 12 ply layup at the fitting, and then carry around the corner of the channel to the flat upper surface. Figures 9-13 and 9-14 show some of these details.

The reason that there are fewer plies as we go forward along the sill is that the load is gradually being picked up by the side of the fuselage as it resists inward bending. There is no exact method to simply calculate how fast this load actually gets transferred to the fuselage shell. We eyeball it, and conclude that, with at least a couple of plies going around and down the outer facing, the load, will get into the fuselage structure with only some large, but acceptable, deflections of the sidewall inward. Of course, as the fuselage sidewall deflects inward, the angle that the seat belt makes will be closer to 90 degrees and the lateral component causing the deflection will, itself, approach zero.

This is typical of the reason that you should always look at the relative stiffness of the structures that you're designing, and be aware of the possible changes in load, and load transference which result from differences between the stiffnesses of different pieces of the structures working together.

The 5-ply figure that this structure begins with in the area of the fitting is calculated by simply using our shear allowable figure, 100 pounds per inch per ply, noting that the 900 pound load will be split between the beam forward and the beam aft of the fitting, and allowing for enough of the 1 inch wide plies to carry the 450 pound load.

The plies going aft from the fitting carry up on to the seat-back bulkhead, just like those on the center console design.

THE CHEST BELT FITTING

The last part of our design will be the anchor point for a cross-chest strap that will restrain the upper body. Let's assume that the lap belt will carry half of the forward load, and all of the vertical load resulting from our 25-G deceleration target. That means the force on the upper fitting will be zero for the upward component, but will be exactly the same as the lap belt fitting for the forward component. This forward component is quite important to restrain as this is what keeps your skull out of the instrument panel during a sudden deceleration. At any rate, the load on the fitting will then be the same 2125 pounds that the lower fittings were designed to take. Again, you should lay out the location of your upper body as it is actually seated in the airplane and carefully locate the point where this attach fitting would best be located. Remember that you want to have the angle of the belt between your shoulder and the upper attach point make an angle of not less than zero degrees, but not more than 30 degrees to a horizontal line. This is because too much download on your shoulder during an impact can cause as much damage as a face full of instruments.

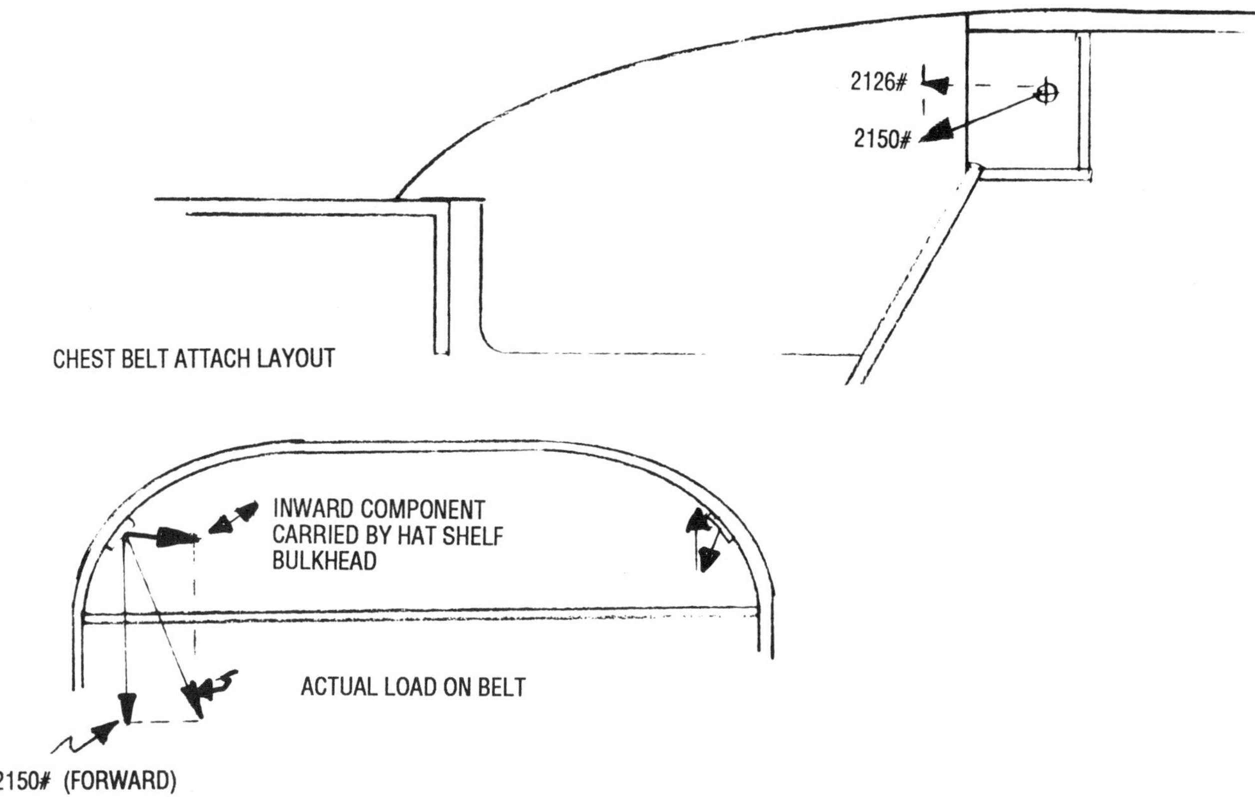

Figure 9-15.

To find the point where the line of the belt will intersect the fuselage shell you must make a carefully drawn side view and top view, to exact scale, of the seating area and the occupants exactly as they will be positioned when properly belted in place in the airplane. For our example problem we will assume that this step has already been completed, and that the intersection point has been found to be as shown in Figure 9-15.

Note that the upper bulkhead, above the storage shelf, is located just aft of the fitting. The reason is that we expect that a substantial lateral load component will need to be supported, and the large cutout made for the canopy opening will leave the fuselage shell in this area incapable of carrying heavy loads in that direction.

The hat-shelf bulkhead is an ideal structure for picking up this load, and also for holding the shape of the same weakened fuselage area under whatever torsional loads might be induced by asymmetrical flight loads. (For this loading condition the hat shelf also becomes primary structure, but the loads will probably be light, so almost any panel we put there will be adequate, as long as the intersections of seat-back, hat-shelf bulkhead, and fuselage shell are all structurally connected.)

The load that our fitting will need to carry will be somewhat larger than the 2125 pound forward load since the belt will be at quite an angle to the forward direction. Rather than try to calculate the value through the trigonometric functions (which you must do to get the exact value), let's lay out a sketch to scale, and scale off the values with a ruler.

If we assume that the drawing in Figure 9-15 is an accurate picture of what we are to build, then the values indicated on the sketch are about what you would get. If we scale the values (lengths) of the components with a ruler, the side view yields a belt load of

2450 pounds, while the aft view makes it look more like 3180 pounds. They are probably both in error, but let's use the larger of the two for sizing the structure that will carry the fitting load. (Incidentally, if you wish to get the right number on a layout that you cannot visualize easily, make a three-dimensional model, and represent the load values with bamboo skewers or piano wire, clipped off to the exact length to represent the load values to scale. This 3-D model can give accurate loads for things like engine mounts, where the angles are difficult to eyeball accurately, and the math is beyond your reliable capability.)

Using the loads indicated in Figure 9-15, the hard-point we are designing will need to carry a forward load of 2125 pounds, a down load of 2250 pounds, and a lateral load of 2250 pounds, so that all of our 25-G conditions can be met with the belt at the angle shown. That hat-shelf bulkhead will come in handy!

For this fitting we will try another way of getting the load transferred from the fitting into the glass structure. We will fabricate an aluminum strap with a bolt hole in it, and bond the strap into the glass structure. For starters, the aluminum we will use, 2024T-3, has the following typical properties:

Tensile strength, Ultimate	68,000 psi
Compressive strength, Ultimate	60,000 psi
Bearing strength, Ultimate	102,000 psi
Shear strength	41,000 psi

The general shape of the aluminum part will be as shown in Figure 9-16, so the critical point in stress will most likely be the hole where the bolt goes through to attach the belt tab. This hole and the bolt will see the full 3180 pound belt load and we calculate the dimensions of the fitting as follows:

Bearing area must be loaded at 102,000 psi, or less, so let's try a 3/8 bolt and a 1/8 thick tab. This gives an area in bearing of 3/8 x 1/8, or 0.0469 square inches. Multiplying this area times the 102,000, we get a load at bearing failure of 4781 pounds, an ample amount, even if we use the usual 1.15 fitting factor on the required 3180 pound load, which would increase that to 3657 pounds. If we use a 1/4 inch bolt, the load capability of the hole would then be 3187 pounds, which is marginal, if we are really going to get the belt load we want. If you think the 3/8 bolt will look too gross, you might try two 3/16 diameter bolts.

The next point is the tear-out strength of the metal near the hole. The tensile strength of our aluminum is 68,000, and we are guess-

View looking down and outboard tangent to fuselage at shoulder belt attach point.

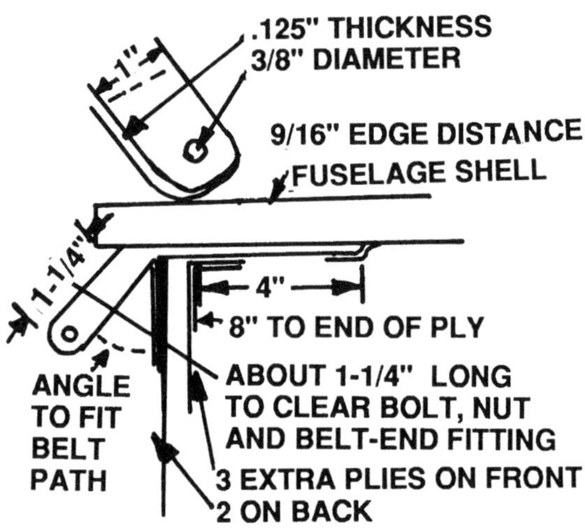

Figure 9-16. Chest Belt Attach Fitting

ing at a thickness of 1/8, so the total edge distance (on both sides of the hole) will need to be the area, divided by 1/8. The area needed is 3180, divided by the allowable stress, 68,000, which comes to 0.0468 square inches. 0.0468 divided by .125 gives us 0.374 inches, or a total of about 3/8 inch, or an edge distance of only 3/16 inch.

However, good practice dictates that we never make a metal part with an edge distance less than 1.5 times the hole's diameter, so we will make the part grossly overstrength, and use an edge distance of 9/16 inch.

With so much margin in both the edge distance and the bearing strength you might want to go back and try the design with a thickness of only .093 inches, instead of 1/8 inch. It will look a little less bulky and might be strong enough to resist casual damage, such as bumping it with your foot. At .093, the allowable bearing load would be about 3557 pounds, almost enough to make it, even when the fitting factor of 1.15 is included!

The part's width can be as little as the hole's diameter, plus two edge distances or 15/16 inch wide. It will look a trifle skimpy at that dimension, however, so let's check the attach area to the fuselage shell to see if it might be better to make it wider.

This area will be fixed by the strength of the bond between the aluminum surface and the glass layers to which it is bonded. Even though the adhesives in common use will yield from 2000 to 5000 psi shear strength, it is good practice to keep this strength level at an allowable for design of about 500 psi. Using this figure we will need enough square inches of bond area in contact with the fuselage shell to generate a total of 2125 pounds load. (Remember, this is the forward component of the load from our original requirements.) The larger force along the line of the belt will be made up of this forward component, along with the contribution of the lateral components being picked up by the hat-shelf bulkhead.

The area we want is found by dividing the load total by the unit load available in the bond, which is 4.25 square inches.

Since both sides of the strip are available to carry load, a one-inch wide strip could be as short as 2 1/8 inches long—except that the glass to which we are gluing it might require a little more area. That area will be determined by the shear strength of the plies along the long edges of the fitting, added to the tensile strength of the glass beyond the end of the strip. (There is no problem with the interlaminar shear strength between plies of the glass since that figure is higher than the shear strength we are using for the adhesive attachment to the plies.)

Using our stress allowables from a previous example, a single ply will carry a shear load of 100 pounds per inch, and the tensile load of 300 pounds per inch. If we use an extra ply under the fitting, next to the single-ply glass facing of the fuselage shell, and a ply over the straight part of the fitting, the three-ply total will carry 300 pounds per inch on each side plus another 900 pounds off the aft end of the fitting. We need a total of 2125 pounds, times a fitting factor of 1.5 (because this whole thing is a composite fitting), which is 3188 pounds. Taking away the 900 we get from the end, we will need a fitting length of 2288, divided by 600 (remember, the fitting has two sides), which is a total length of 3.81 inches. Let's just settle for four inches long.

Now let's calculate how far these plies must extend into the fuselage shell. Both facings will pick up the load, but will need a few inches more to do it. Drop off the ply under the fitting about an inch away in all directions, and then extend the top ply another inch to the sides, and about four more inches to the rear.

The extra plies needed on the hat-shelf bulkhead, where the fitting will be pressing inward with the 2250 pound side load, will need a bearing area of 2250, divided by 50,000, which is the bearing strength allowable of the glass. This makes the needed area equal to 0.045 square inches, and since the fitting is one inch wide, that is also the thickness total required. Using our 1.5 composite fitting factor, this comes to 0.067 inches total, or seven plies. There are already two plies carrying a facing load in edgewise compression. The facing compressive allowable, established earlier is 250 pounds per inch, per ply, or 500 pounds per inch total for the two plies. The last ply must, therefore, engage a total width of the panel of 2250 divided by 500, or 4.5 inches. The same 1.5 factor makes this 6.75 inches, and common practice requires this dimension for both the length and the width of the area. In other words, draw a 6.75 inch radius from the middle of the contact point, and use this for the outline of the top ply.

One other point should be kept in mind when using bonded-in metal fittings. Unless you properly treat the metal for bonding, the interface between the metal and the resin will delaminate in time, (sometimes a **very** short time) and all of these fine calculations will be worthless.

It does no good to use even lower allowables because the nature of the delamination is such that it continues to progress until the parts are completely separated. You only buy another month, up to as much as a year or two, before you have a non-structure on your hands. If you are to use this attractive method of design, you **MUST** be sure to have the aluminum Phosphoric Acid Anodized, primed with Corrosion Inhibiting Primer, in careful conformance with any one of the specifications developed by one of the major airframe manufacturers, such as Boeing Specification BAC-5555. (The other major airframe makers have similar specifications.) This process is covered in Chapter 11 and Appendix 4.

SUMMARY

To review some of the things that have been considered and solved, we can make the following observations:

1. The design job cannot start until we have decided exactly what materials will be used, and what strength these materials will give us in the planned application.
2. Look out for the problem of attaching a new structure to an existing structure. It is always advisable to make a simple test of the materials that you plan to use in order to determine that the strengths you anticipate will actually be developed.
3. Do not become too entranced with the local piece of hardware that you are designing at the moment, as it may not fit into the scheme of things when you look at the entire structural job that must be accomplished overall.

4. When adding many plies of fabric for a local strong-point, such as the area in the immediate vicinity of the bushing in our design, make the transition to the lighter structure by dropping off one ply at a time, **at least 1/2 inch apart**, so that the transition is nice and gradual. Large loads do **not** like a sudden change in cross-section!
5. Do not assume, without a careful review of both the strength of the structure **AND** its relative stiffness, that the structure to which you attach your load will carry that load.
6. Both sides of a sandwich can pick up a load put into one side, but the core can only transfer load in an amount equal to its shear strength, multiplied by its area loaded. Urethane foam at a 4 to 5 pcf density, for example, needs 1 square inch of sandwich for each 40 pounds of load to be transferred. Transferring half of a 2000 pound load from the front facing to the rear facing would take at least 25 square inches of panel.
7. Conduct some strength tests, particularly if you plan to attach a new composite structure to an existing structure. Don't be afraid to jury-rig some simple load tests. Seat belts, attachments and all, can easily be tested by running a couple of extra long belts over a sturdy beam, like a 6 x 6, and then hoisting up the whole airplane by that beam. With a few extra friends draped over the cowl and the aft fuselage, you can generate substantial loads and verify your seat belt integrity. (The forward load is a little more complicated!)
8. Don't assume that some particular design is best simply because you have seen it in somebody else's structure. The person who designed that fitting has probably already moved on to better designs. After all, just see how the fittings in this Chapter improved on the third try. That shoulder harness fitting looks a lot simpler and lighter than either of the lap belt fittings!
9. If you wish to do a better job with the design of your structure, buy one of the PC programs on composite sandwich and laminate design. They are very easy to use, and will forever keep your arithmetic straight. One such program is,"PBJ", available from Antrim Associates at the address listed in Appendix 1.
10. **DO NOT** use the allowables that were used in this book on the example problems. They came out of thin air, and may, or may not, represent something close to what your materials and methods will deliver on your project.

PERFORM STRENGTH TESTS ON YOUR OWN PARTS! THAT IS THE ONLY WAY YOU CAN BE SURE EXACTLY WHAT LOAD THEY WILL ACTUALLY CARRY!

CHAPTER 10

CALCULATING STRESSES AND LOADS

In the previous chapter it became obvious that the design function cannot logically proceed unless the designer has an understanding of the loads that will be imposed upon the structure. In addition, we must understand how the structure will pick up these loads, and how they will be carried to some point, where they will be transferred to another piece of structure, to the ground, to the air, to the water, or to whatever is the end force that resists the original load. In order to ensure that the part being designed is strong enough, but does not have any unnecessary strength (and weight), these loads must be reduced to exact numbers and then matched to structural elements that are the right shape and size to carry them. Many people avoid becoming involved in these calculations because they are not engineers and feel that the mathematics are too complex. Actually, most of the calculations are quite simple and can be reliably performed by anyone who has eighth grade arithmetic skills. This chapter will provide some needed and simple engineering background so that you can do a large proportion of the load and stress work without the use of higher mathematics or computers.

Engineers and structural analysts are much like doctors in that they have developed a special language for use within their field of technology, and those outside of the field are often excluded by this "language barrier." Some of the important terms are as follows:

1. LOAD is the gross or total amount of force that is applied to something, usually expressed in pounds. A load always has an amount (for example, 100 pounds), and a direction, which is usually indicated by an arrow, as shown in Figure 10-1. Every load also produces an equal and opposite load, called the reaction, which keeps the entire system in balance, or at rest. (That is why this sort of analysis is called a "static" analysis.) The reaction will often be left out or ignored when sketching a loading arrangement, but it is always there and must be considered, especially if it is located in another part of the structure that you are also designing.

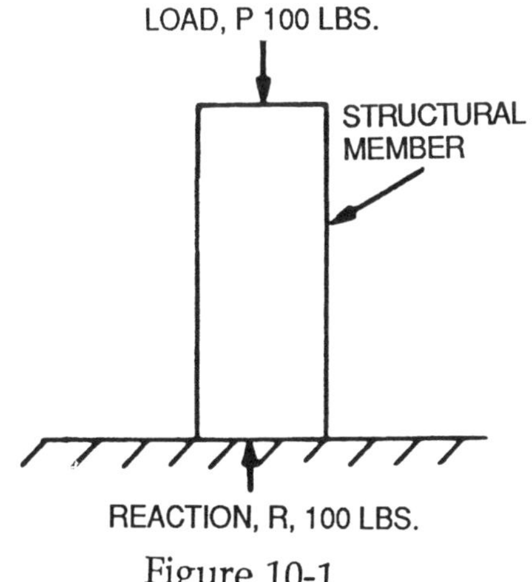

Figure 10-1.

2. WING LOADING is the amount of load applied per unit of area of the wing, such as "10 pounds per square foot of wing area." Also used for horizontal and vertical tail areas. This is not to be confused with the load on a structural member, as it is the air load on the surface that is the source of the flight loads on various individual members in the structure.

3. STRESS, also called "unit stress," this is the amount of load applied per unit of area, but in this case, the area is the cross-sectional area of the structural member that carries the load. For example, if a 1/2 inch x 1 inch bar of material is loaded in tension with a load of 100 pounds, as shown in Figure 10-2, the stress would be 200 pounds per square inch, since only half of 1 square inch of cross-sectional area is carrying the 100 pound load. (The length of the 1/2 x 1 bar does not affect this figure.)

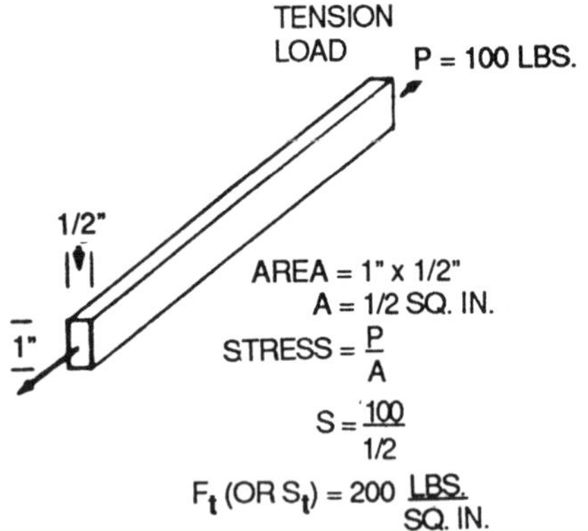

Figure 10-2.

4. ALLOWABLE STRESS is the exact number of pounds of load that can be carried by a piece of material having 1 square inch of cross-section without failing. In practice, this number is chosen to be slightly below the expected load at failure, in order to allow for a reasonable variation in the strength of different pieces of the same material. In the case of composite materials, particularly those produced in a home workshop environment, this difference between the average strength of all parts produced and the actual measured strength of the lowest-strength piece in the lot, can be as much as 20% to 30% of the average value. "Allowables" for homebuilt parts are, therefore, usually lower than allowables for nearly identical parts that are made under factory controlled conditions. The last part of this chapter deals with this problem in more detail.

5. LOAD DIRECTION is as important as the load itself, since most of the structural materials that are used in real structures (especially composites), are stronger in one direction than they are in another.

6. STRENGTH DIRECTION is the exact direction within a structural material in which a quoted amount of allowable stress value may be carried. When a composite is made of separate layers of fibers, in which each layer is oriented in a particular direction, the entire piece of material will have a different allowable stress in each of several directions. Be careful, because the strength in a direction between two quoted directions is usually lower than either of the quoted strength figures, and a typical two-directional structure of a woven fabric skin material will have a strong direction, in which most of the fibers are oriented, and a weak direction, which is the direction of the cross-fibers, and an even weaker strength halfway between these two directions.

ONE OF THE PRIMARY ADVANTAGES OF COMPOSITES IS THAT THE STRENGTH IN SEVERAL DIFFERENT DIRECTIONS CAN BE TAILORED TO MATCH THE EXPECTED LOADS IN EACH OF THOSE DIRECTIONS.

7. STRENGTH is the total load that a given structural member can carry in a given direction. A high-strength material can carry the same load as the much thicker section of a low-strength material.

8. STRAIN is the amount that a loaded structural member deflects due to the load it is carrying, as shown in Figure 10-3. All structural materials deflect under load and the exact amount that they will deflect is measured by their "modulus."

9. MODULUS is a precise measurement of exactly how much a material will stretch under a given load. A high modulus means that the structure does not move much. You cannot see the amount of stretch in a steel bolt when the nut is tightened (unless you use a micrometer), but if the bolt were made of rubber, which is a low-modulus material, you would see a substantial increase in its length when tightened to the same nut torque. The figure used for modulus is literally the unit stress in pounds per square inch, which would make a piece of the material stretch elastically to twice its original length.

Of course, most materials are not strong enough to double their length under load without either yielding and deforming, or breaking. However, this figure can be used at the actual loads that are carried in order to calculate the exact change in the part's dimensions. Using virtually this same approach, the designers of the Boeing 777 Transport will calculate the exact deflections at each point in the wing in their static test program, and will miss by only a percent or two.

10. TENSION is a load acting along a single line, which tries to stretch the member that is being loaded, as shown in Figure 10-2.

11. COMPRESSION is a load that acts along a single line, or at right angles against a surface, trying to shorten, or squeeze the member under load, as shown in Figure 10-4.

12. SHEAR is a load that acts parallel and in the opposite direction to an equal resisting force off to one side, as shown in Figure 10-4.

13. BEARING is the special case of a compression load that a small member, such as a bolt, imposes upon the hole in which it is located when the load is at a right angle to the axis of the bolt, as shown in Figure 10-5.

14. BENDING is a complex load that usually includes all of the above three loads as components. Figure 10-4 shows the simple cantilever (supported only at one end) beam of Figure 10-3, but with the load components as they would appear near the left, or fixed end of the beam. Although the shear is shown at the fixed end of the beam, it would be the same value at each point, all the way out to the loading point. The compression and tension values are a maximum at the fixed end and drop toward zero as

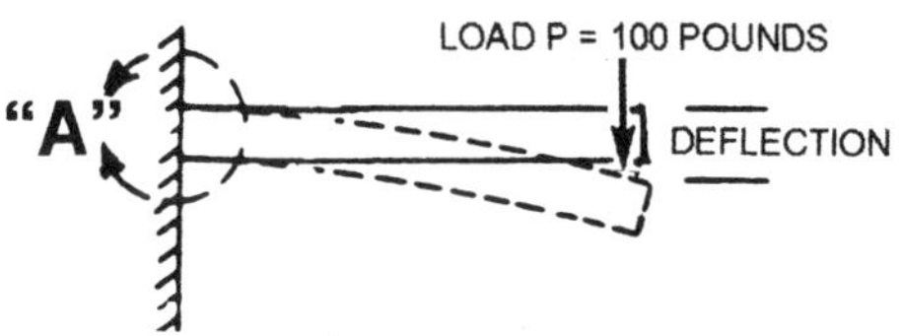

"DEFLECTION" IS USED TO DESCRIBE THE TOTAL MOVEMENT OF THE MEMBER. "STRAIN" IS THE DEFLECTION OF A UNIT OF AREA PER UNHIT OF APPLIED LOAD. BE CAREFUL, THOUGH, AS MANY VARIATIONS OF THESE TERMS ARE IN COMMON USE!

Figure 10-3.

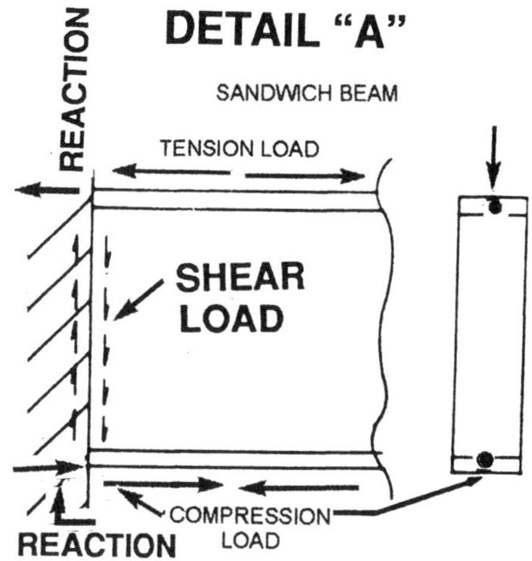

IN A SANDWICH BEAM THE CAPS ARE ASSUMED TO CARRY **ALL** THE TENSION AND COMPRESSION LOADS, AND THE WEB TO CARRY ALL OF THE SHEAR LOADS.

NOTE: THE ● IN THE END VIEW OF THE SPAR CAPS REPRESENTS THE END VIEW OF AN ARROW.

Figure 10-4.

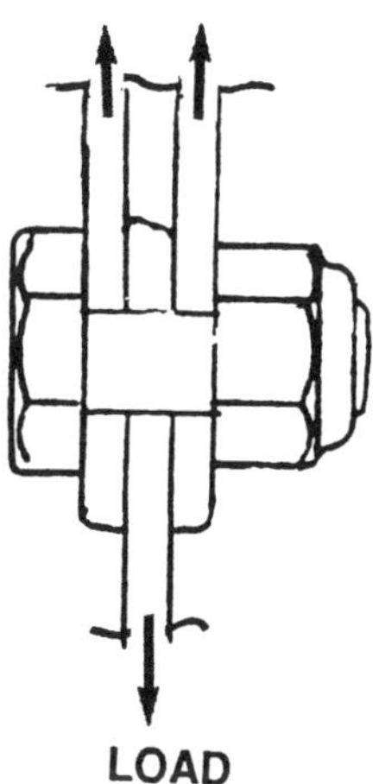

Figure 10-5. In a bolted or riveted connection the tension load on the members results in a bearing load on the inner surface of the hole.

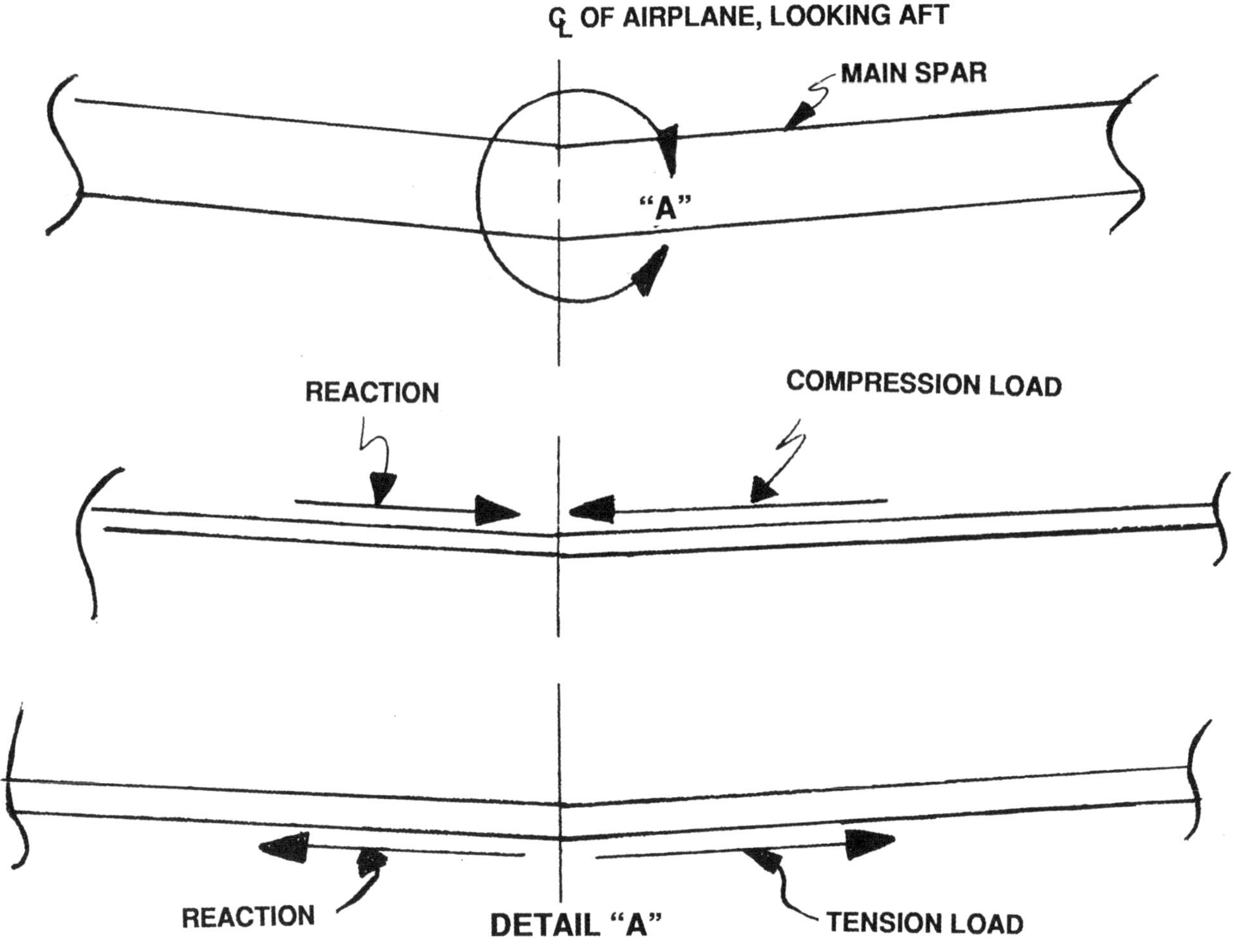

Figure 10-6. Wing root reaction forces.

the point that is being checked moves out toward the loading point. Even though the sketch looks like an unlikely situation, it almost exactly represents a wing spar in which the end fixity is provided at the fuselage attach-point by the other half of the wing, as shown in Figure 10-6. In this example, the shear load, P/2 is all reacted into the fuselage attach fitting at wing station 12, as shown in Figure 10-7.

15. LOAD COMPONENT: When a load is in an odd direction, we sometimes separate it into two or more components, each of which is acting in a direction that is easier to analyze, or one which lies along a member which will carry one of the reaction forces. The sum of the components is equal (geometrically, not arithmetically), to the original load, as shown in Figure 10-7.

16. LOAD RESULTANT: When more than one load acts upon a part at the same time, we can add them geometrically as shown in Figure 10-8, but in reverse, to get a single load. This single load can then be separated into components that are easier to analyze, or that fit the structure better.

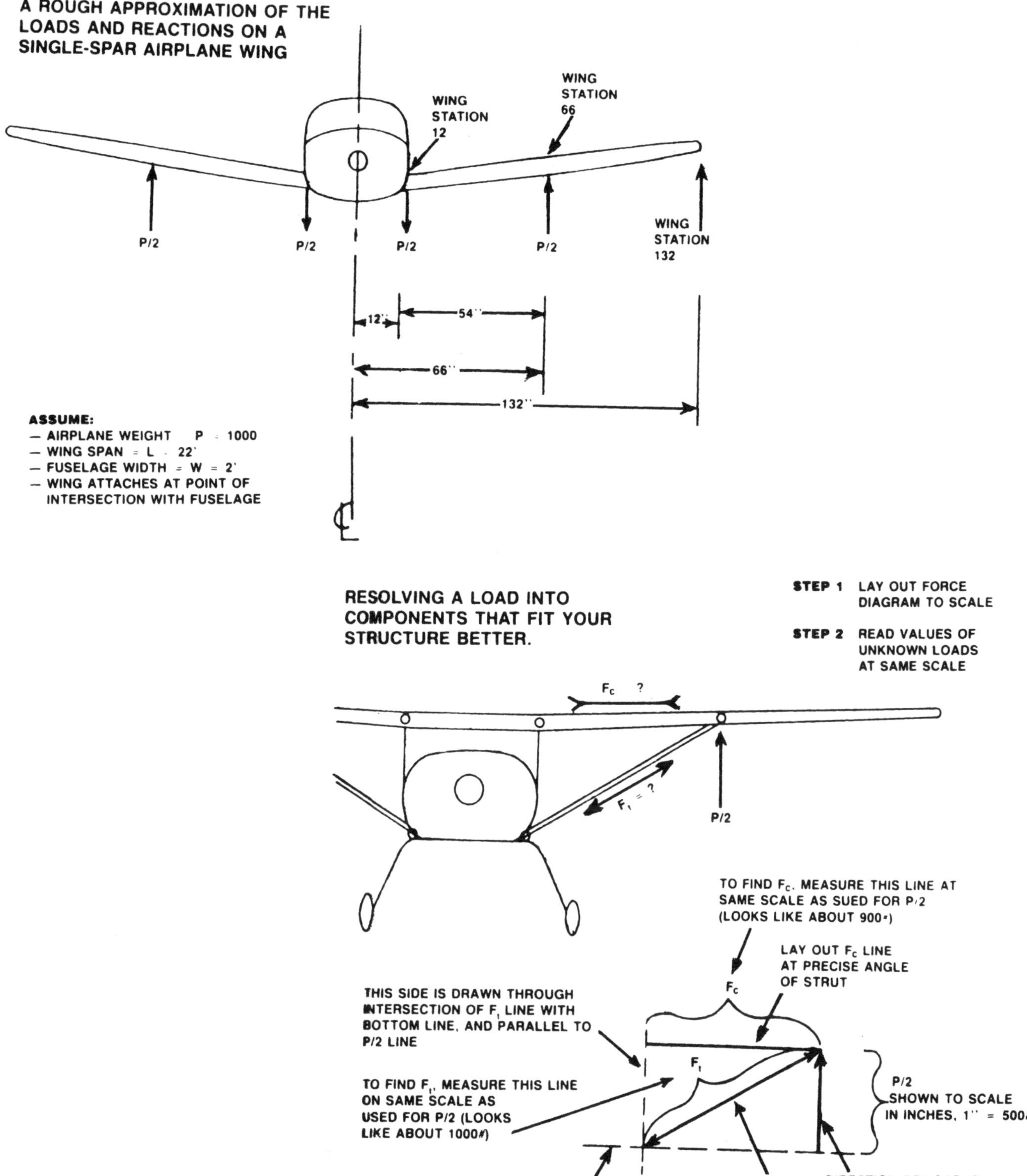
A ROUGH APPROXIMATION OF THE LOADS AND REACTIONS ON A SINGLE-SPAR AIRPLANE WING
WING STATION 12
WING STATION 66
WING STATION 132
P/2
P/2
P/2
P/2
12"
54"
66"
132"
ASSUME:
— AIRPLANE WEIGHT P = 1000
— WING SPAN = L = 22'
— FUSELAGE WIDTH = W = 2'
— WING ATTACHES AT POINT OF INTERSECTION WITH FUSELAGE
RESOLVING A LOAD INTO COMPONENTS THAT FIT YOUR STRUCTURE BETTER.
STEP 1 LAY OUT FORCE DIAGRAM TO SCALE
STEP 2 READ VALUES OF UNKNOWN LOADS AT SAME SCALE
Fc ?
Ft = ?
P/2
TO FIND Fc, MEASURE THIS LINE AT SAME SCALE AS SUED FOR P/2 (LOOKS LIKE ABOUT 900#)
LAY OUT Fc LINE AT PRECISE ANGLE OF STRUT
Fc
THIS SIDE IS DRAWN THROUGH INTERSECTION OF Ft LINE WITH BOTTOM LINE, AND PARALLEL TO P/2 LINE
Ft
TO FIND Ft, MEASURE THIS LINE ON SAME SCALE AS USED FOR P/2 (LOOKS LIKE ABOUT 1000#)
P/2 SHOWN TO SCALE IN INCHES, 1" = 500#
LAY OUT Ft LINE AT PRECISE ANGLE OF STRUT
DIRECTION OF LOAD IS AT EXACTLY 90° TO WING CHORD PLANE
THIS SIDE IS DRAWN THROUGH END OF P/2 ARROW, PARALLEL TO Fc LINE
NOTE: SUMMATION OF ALL LOADS AT POINT ANALYZED MUST EQUAL ZERO

Figure 10-7.

DEFINITION: MOMENT = FORCE (OR LOAD) TIMES DISTANCE FROM POINT AT WHICH MOMENT IS TO BE FOUND.

ALWAYS TRUE: TOTAL OF ALL MOMENTS ABOUT ANY POINT IN THE STRUCTURE IS ALWAYS ZERO

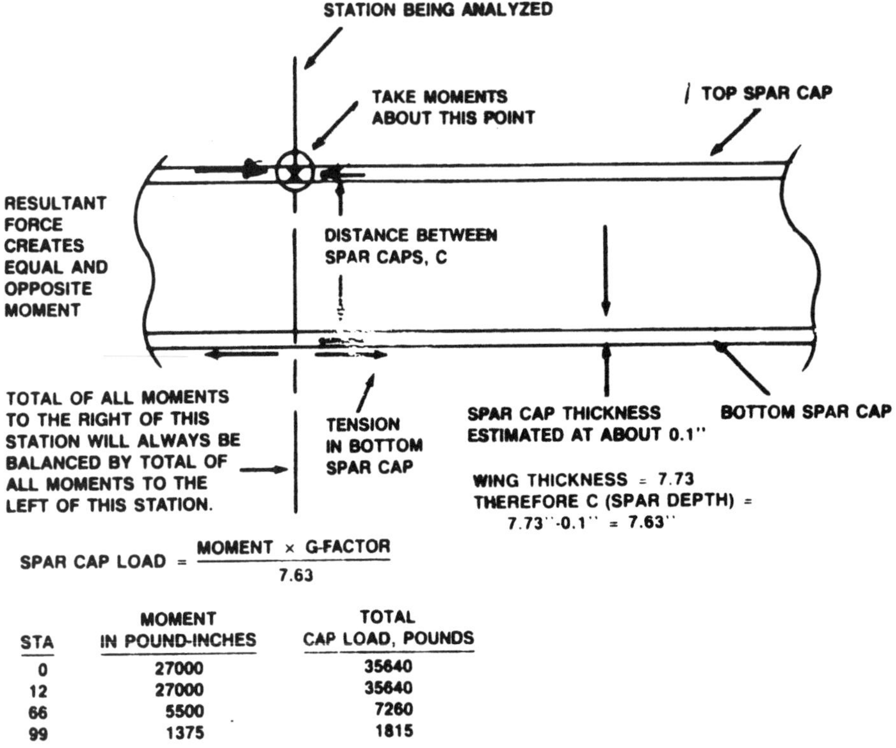

$$\text{SPAR CAP LOAD} = \frac{\text{MOMENT} \times \text{G-FACTOR}}{7.63}$$

STA	MOMENT IN POUND-INCHES	TOTAL CAP LOAD, POUNDS
0	27000	35640
12	27000	35640
66	5500	7260
99	1375	1815

Figure 10-8.

The terms "Tension, Compression, Bending, Bearing," and "Shear," are used to describe forces, stresses or strength, as well as to describe loads. For example, many materials have a different strength in compression than they have in tension. Virtually all materials have a much different strength in shear than in either tension or compression. Consequently, it is always a good idea to check the strength of the material that you are using in order to find its capability in each of the types of loading to which the part will be subjected.

In addition to the terms explained above, many other types of loading exist and must be analyzed. Most of them, however, can be broken down into the Basic Three Loads, -- Tension, Compression and Shear, -- or they can be approximated for a preliminary design effort. This chapter will not attempt to describe the more complicated types of analysis, and will offer only a few shortcuts and "quick estimate" approaches instead, at some sacrifice of accuracy.

IF YOU WANT A DETAILED AND COMPREHENSIVE STRESS ANALYSIS, YOU MUST GO MUCH FARTHER THAN THE METHODS OUTLINED HERE.

WHAT TO DO FIRST

Before performing a stress analysis, we must design the structure, and before we can even begin this, we must perform a "load analysis." In a typical airplane design, this load analysis consists of writing down all of the operating conditions under which the airplane must perform, and then examining all of the key elements of the structure to see where the loads might be critical. For example, some of these conditions might be:

CONDITION 1: Starting, stopping, taxi, takeoff roll and landing runout. -- Critical points are:

1. Torque on engine mount and carrythrough structure caused by back-firing during starting.
2. Rearward load in wheels due to striking a rut or small crosswise ditch.
3. Download on nose wheel due to maximum braking or the ditch load of item 2 above, on main wheels.
4. And various others.

CONDITION 2: Hard landing at maximum gross weight. -- Critical points are:

1. Downward load from full fuel tanks.
2. Upload load from gear legs into gear attach points.
3. Same as above, but for nose-down landing.
4. Rearward load from set brakes, combined with maximum up-load on gear fully flattened.
6. Side loads from cross-wind landing on one wheel.
7. Side loads from ground loop.
8. And various others.

CONDITION 3: High-speed flight at maximum gross. -- Critical points:

1. Wing-bending strength at centerline of the fuselage, fuselage attachment to wing spar, gear attachment, strut attachment.
2. Wing attach and carry-through structure.
3. Canopy attach strength.
4. Canopy deflection.
5. Horizontal tail-bending strength at maximum trim and elevator deflection, both up and down.
6. Vertical tail-bending strength at maximum rudder deflection.
7. Wing drag strength.
8. And various others.

CONDITION 4: High speed at flaps and gear-down condition at maximum gross weight. -- Critical points are:

Same as Condition 3.

A listing of all of the above items, plus many not included in the above example, is important, since some of the "worst conditions" for design of some particular point in the structure may occur within some unsuspected part of your performance envelope. Landing gear structures in particular may be more sensitive to failure from taxi through a small ditch than from rather hard landings. Slender tail booms may reach maximum loading under gusty winds while tied down at the airport.

A complete list is nearly endless, but a good place to start is to go through those sections of the Federal Air Regulations that govern the structural requirements of certificated aircraft of the type closest to the one that you are designing. At least then all of the commonly known design conditions will be covered, and you need only concern yourself with those areas in which your design is unique.

After you have completed this list of critical design conditions and located the corresponding critical points in the structure, almost all of your analysis efforts will be directed toward these critical points.

WHAT G-LOADING TO USE

All of the examples we've used so far have referred to the gross weight of the airplane as the basic load figure from which many other specific loads are derived. Because real airplanes are subjected to acceleration forces of several times their own weight in many maneuvers, and are also subjected to wind gusts and turbulence, it is common to apply a factor of from 4.5 to 10 or more to the basic derived loads. Much of the structure that you will build is going to end up overstrength simply because the flight loads are so low that the corresponding skin thicknesses, core densities and laminates will then fail under the normal handling, pushing, bumping and jostling that all airplanes experience. They will, therefore, be up-graded from the calculated thicknesses to something that you judge will withstand such "garbage" loads.

The parts for which flight loads will dictate the weights and thicknesses used are usually spar webs and caps, attach fittings, gear legs and a few others. These represent such a small part of the airplane weight that changing from an absolute minimum of 4.5 G's or so, up to a more reassuring 9 or 10 G's usually results in a rather minor increase in airframe weight. Along with this feeling of reassurance, you will also buy a relative immunity from such problems as minor errors in resin-hardener ratio, poor layup techniques, normal aging and long-term structural deterioration, errors in stress calculation and structural testing, and many of the other worries that plague anyone who begins a project from scratch. A minimum load factor of at least 6 is therefore strongly recommended for even non-aerobatic aircraft designs.

SOME SPECIFIC STRENGTH CALCULATIONS

Having found the key points of our structure, let's examine just how to calculate the loads to be carried by each element.

WING SPAR: Referring again to Figure 10-6 and assuming that we will design our airplane to withstand 10 G's, let's look at some specifics. The depth of the spar will be about equal to the thickness of our airfoil, so we will need additional data in order to calculate this depth. At a wing loading of 12 pounds per square foot, (a typical figure for small aircraft) our wing area will be 1000/12, or 83.3 square feet, and, with a straight rectangular wing planform, the chord will be 83.3/22, or 3.78 feet. 3.78 x 12 is a chord of about 45.5 inches. If we also assume that we will use a NASA LS(1)-0417(MOD) airfoil, one of the common current choices, the airfoil will be 17% thick, or .17 x 45.5 = 7.73 inches thick.

This thickness will be the same from root to tip, provided that we use a rectangular planform wing as mentioned above, and the same airfoil for the full length. For other choices, you use the same calculations at several points along the spar in order to get the spar depth at these points for a tapered thickness or for a tapered wing planform, which automatically gives you a tapered spar because of the airfoil section's smaller chord. The dimension of the spar is critical in determining the spar cap loads, so all of the wing planform and airfoil choices must be made **before** you begin the stress calculations.

Again referring to Figure 10-6, let's calculate the bending moment at wing stations 0, 12 and 66. This is a static analysis, so we assume that the loads and reactions of the left wing are the same as those on the right wing. In other words, each half thinks it is simply stuck on to an infinitely stiff support like that shown in Figure 10-3. The difference is that in our "real airplane" case, the up-load at station 66 is exactly balanced by the down-load at station 12. The opposite wing's loads balance out these loads exactly, so we know that:

1. All of the stresses in half of the wing, beginning at the centerline and extending to the tip, will be the same as those in the opposite wing half. We therefore only need to look at one side.
2. The air load on the wing-half, P/2, is matched exactly by the reaction at the station 12 fuselage attach point, so that the shear in the spar drops to zero between station 12 on the left and station 12 on the right.

We also know from the laws of physics that the sum of the moments about any body in equilibrium must total to zero. We can therefore derive the bending moment, M, at any point on the spar by totaling the products of each load, multiplied by its distance from the point that is being considered. We know that the total to the right of our point will be equal to the total to the left of our point, but are opposite signs. (Assume that a "+" denotes the counter-clockwise moments, and a "-" denotes the clockwise moments.)

Taking all of the above on faith, the bending moment at station zero will be M = [12 x (-P/2)] + [66 x (+P/2)] and reducing that to numbers, the first quantity is a -6000, the second quantity is a +33,000, and the total moment then equals +27,000 inch-pounds. Calculating in the same manner at station 12, the moments equal [66 x (+P/2)], which equals +27,000 inch-pounds.

If we really believed that the air loads were acting at a single point on the wing at station 66, then there would be no spar load at all outboard of that point. We know that this is not true, so let's just assume that half of this load, or P/4, is acting at a point that is one-third of the way from station 66 to station 132, or station 88. Therefore, as an approximation, at station 66 the moment is equal to [22 x (+P/4)], which is equal to 5500.

If we want to check another point half-way between station 66 and the end of the wing, station 99, we could do the same thing, only the load would be P/8, and would be about one-third of the way from station 99 to station 132, or at station 110. At station 99, the moment equals [11 x (+P/8)], which is equal to +1375 inch-pounds.

We use a distance of one-third the way from the station that is checked to the wing tip as the point where the remaining load is acting because the load tapers off and drops to zero at the tip. This assumption is not accurate, but it's much quicker than the "more accurate" methods, and usually introduces only minor errors. Even when really good data and analysis methods are used, the bending strength of the final product must always be checked by static loading to an acceptable level (not necessarily to failure).

Our "quick and dirty" analysis has now given us the approximate value of M at stations 0, 12, 66 and 99. These values will be used to determine just how much bending material must be placed in the cap areas of the spar, where the bending loads are carried. The formula normally used is: S = MC/I, or, Stress = Moment x one half the Spar Depth, divided by the section's moment of inertia. For a quicker answer, however, we can estimate the spar depth as being about one spar-cap thickness less than the airfoil thickness, and simply forget about I, moment of inertia. We do this by taking moments about one spar cap, as seen in Figure 10-8. For our example we have figured that the total wing thickness is 7.73 inches and guessed at a cap thickness of 0.100, so C will be about 7.63 inches.

ALLOWABLE STRESS LEVELS

In order to calculate the spar-cap thickness (or, more accurately, the cross-sectional area of spar-cap material) at each station that is being examined, we must discuss the allowable stress of the spar-cap material. In Figure 10-9 below are realistic estimates you might use.

STRENGTH ESTIMATES FOR SEVERAL UNIDIRECTIONAL COMPOSITES

MATERIAL	TENSION ALLOWABLE PSI	COMPRESSION ALLOWABLE PSI
Hand-laid, unidirectional E-glass fiber in polyester resin	40,000	35,000
Hand-laid unidirectional E-glass fiber in vinylester resin	50,000	40,000
Hand-laid unidirectional E-glass fiber in room temperature cured epoxy	50,000	40,000
Vacuum-bagged, prepreg unidirectional E-glass in epoxy, 250° F cure	120,000	110,000
Precured E-glass laminate, unidirectional, 28 % resin, 350° F cure at 100 psi ("Spar-Tuf")	140,000	150,000
S-glass unidirectional	Add 20% to E-glass tension valuesonly.	
Carbon fiber unidirectional	Same as E-glass at low end, add 80% at high end.	
KEVLAR 49 UNIDIRECTIONAL		
With polyesters	45,000	35,000
With vinyl esters	70,000	40,000
With epoxy hand-layup	80,000	45,000
250-degree vac-bag, prepreg	150,000	70,000

NOTE:

(1) It is quite probable that NONE of the above figures will be duplicated with your specific resin system and process. They are only offered as a "first guess" guide.

(2) Hand laid carbon fiber rarely exceeds 50,000 tensile or 40,000 compressive strength. Prepreg can be little better unless tooling is used to insure straightness and low resin content. However, precured pultruded material may exhibit 4 to 8 times these figures. (See Page 10-17.)

You must make and test actual samples using your own materials and processes in order to find the stress values that you can expect, as well as the variation in strength from one batch to another of the structural parts that are actually produced.

Figure 10-9

SPAR-CAP DIMENSIONS

Again refer to Figure 10-8 to estimate the value of C, the spar depth, as 7.63 inches. If we again use the sum of the moments about a point as equal to zero, the total tension load in the bottom cap, at station zero or at station 12, would be T, and this value x 7.63 would equal 27,000 inch-pounds, or more simply, T = 27,000 inch-pounds divided by 7.63 inches, which equals 3539 pounds. This load could be carried by 1 square inch of a material that has a strength of 3539 psi, (let's use 3500 psi for a simplified example), or half of 1 square inch of something that has a strength of 7000 psi, or 1/4 square inch of something that has a strength of 14,000 psi, or 1/40 square inch of something that has a strength of 140,000 psi. 1/40 of a square inch would be 1 inch wide by 0.025 inches thick.

All of the above only consider the 1 G load figures. At 10 G's the values would be multiplied by 10, so the .025 sq. inch of spar cap becomes a more realistic 0.250, or 1/4 sq. inch. Since we have a constant-depth spar, the spar cap areas for the other stations are in direct proportion to the moment at that station.

THE SPAR WEB

In either a sandwich spar, such as the one we are analyzing, a tension field beam like those seen on most aluminum airplanes, or on the hybrid structure that is favored by Burt Rutan, the spar caps are assumed to carry all of the bending loads by exerting a tension and compression force in the bottom and top caps respectively, to exactly equal the bending moment that is created in the spar by the air loads.

These spar arrangements also assume that the spar web is carrying all, or almost all, of the shear loads. The shear loads are much easier to figure, since at any point they are equal to the net of all of the up-and-down loads between that point and the end of the wing. Even though the spar web loads are easier to calculate, the stress allowables for the materials that you might use are just as difficult to pin down. As a first approximation, the shear strength of a composite web can be assumed to be about half its tensile or compressive strength. Looking back at our table of suggested compressive and tensile allowables does not lend much comfort. Again, it must be emphasized that actual parts should be tested to failure if any real reliability is to be expected from the chosen values and the calculated strengths.

For our example of Figure 10-6, the shear load would be zero at the wing tip, then gradually build to a maximum at wing station 12, where it goes back to zero. The conservative assumption is that the shear load varies directly with the distance from the tip to station 12. The value at station 12 is equal to P/2 times the G-loading chosen, or 5000 pounds. If a sandwich spar web is used, the full 7.73 inches deep, and two equal facings on a foam or honeycomb core of about 1/4 inch thick, a total depth of 15.46 inches of sandwich facings will carry the load. 5000 pounds spread out over 15.46 inches means that each inch of the facing must carry 5000 divided by 15.46, or 323.5 pounds per inch of facing.

As we mentioned earlier, the shear strength of many composite laminates is only about half of their tensile strength or compressive strength. In a spar web, however, we have a special condition. The web shown in Figure 10-10 is the way it would look if an engineer who was accustomed to aluminum were to draw in the load arrows. The load P is acting down, so the little half-arrows that are usually used to depict shear loads are drawn parallel to the main load arrow; no attention would be paid to the orientation of the yarn in the fabric weave that is used for the web. However, if, as shown in Figure 10-11, we use a different method to signify the shear loads and spar deflection, and then split the shear load arrows into two components, every filament in the spar web fabric will be equally loaded, the warp yarns in tension, and the fill yarns in compression. As a result, the resistance by the rectangular section of spar to becoming a parallelogram will be substantially higher.

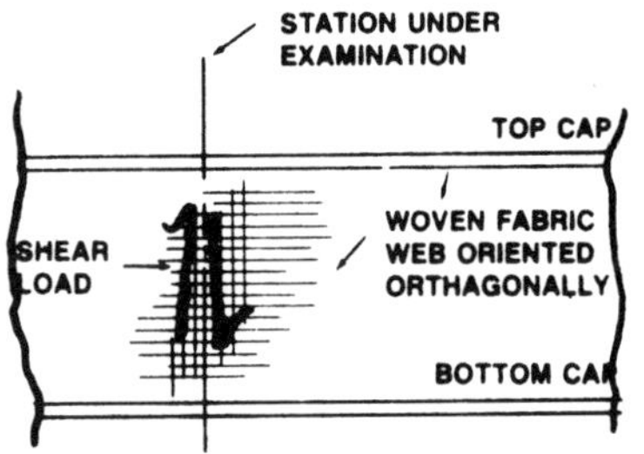

Figure 10-10

If the geometry in the force diagram is assumed to be correct, (view C of Figure 10-11), then the load that the fabric layer can carry in shear will be the same as that which it carries in tension or compression (provided that it is a balanced weave which has the same yarn count and yarn construction in the warp direction as in the fill direction). Since any fabric supplier can provide the figures for fabric tensile strength in both directions, you have only to choose one that exceeds the calculated value, 323.5 pounds per inch of width, in each direction. To allow for normal variations in layup orientation and such practical problems, this strength is reduced by a factor of 2. The chosen fabric must then be oriented with the yarns running at 45 degrees to the spar length. To taper the spar web weight and strength, more than one layer of a weaker fabric can be used, dropping off a layer at appropriate points, as the structure toward the tip requires a lower number of pounds per inch of spar depth.

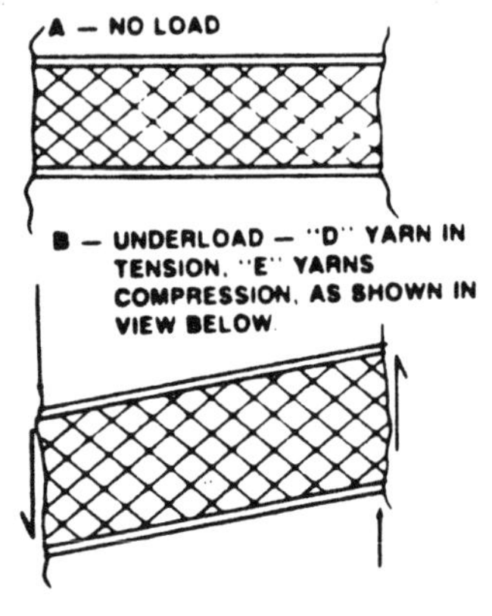

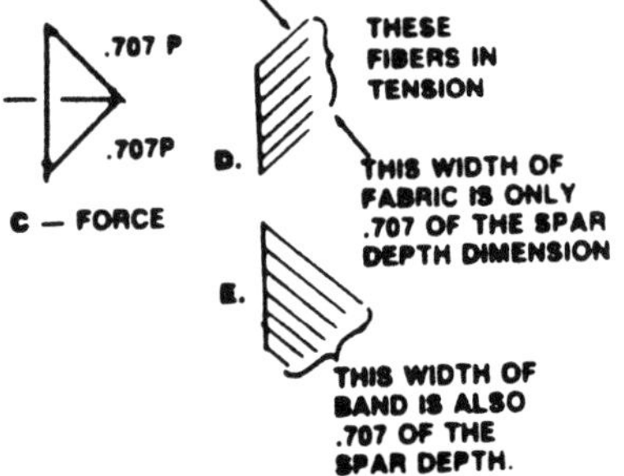

Figure 10-11

TAIL STRUCTURE

The horizontal and vertical tail structure is treated much the same as the wing structure, except that the loads are smaller, and the structure can therefore be somewhat lighter.

HINGE AND ACTUATOR LOADS

Both the wing and tail surfaces will need a fitting or strong point at which to pick up the load at hinges, bellcranks and tie- down rings. In order to carry these loads, one first draws a diagram that shows the direction or directions the load will be traveling, and then calculates or estimates the number of pounds, including G factor, that you will need to carry them. As shown in Figure 10-12, these loads are often in a location and direction where no existing structure can be used to pick them up. We simply find a nearby piece of structure and provide a path that consists of suitable pieces of material so that the load can follow this material into the existing structure. Do not be constrained to follow the suggested design, as it is simply an example of one of the ways in which the job might be accomplished.

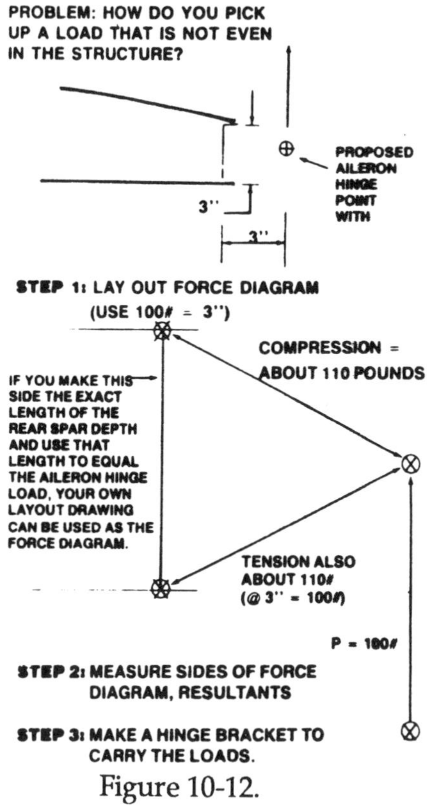

Figure 10-12.

DESIGN VERIFICATION

Readers who have managed to follow this Chapter to this point are probably somewhat disappointed in the lack of specific strength figures offered, and the rather loose way in which structural analysis has been treated. I share your feeling, but suggest physical testing. If you are to have a lightweight structure that is capable of exceeding your target strength, you must load it with the expected loads to verify your design and construction process. If the job is approached carefully, most of the major structural elements will be usable after the tests have been completed, and the minor elements can be tested as single samples that represent many other items which are used without individually testing each one. The different test methods that can be used in a home workshop are varied enough to be the subject of a separate chapter, but be assured that you will definitely not know what the real strength of your structure is if these tests are not conducted.

Whether you use a $200,000 Instron Testing Machine to produce the test loads, or, as a low-cost alternative, use sandbags, hydraulic jacks, weights and levers, it is absolutely necessary that the tests be conducted. In addition, because the actual strength of the parts you build is so dependent upon the details of the way they are assembled and cured, the need for testing is just as important for a design which has been checked by a professional stress analyst as it is for parts designed by the simple analysis covered here.

WHAT ALLOWABLES ARE CORRECT?

In the earlier portions of this chapter, many specific values have been used in examples where a "design allowable" was needed for a specific part. Most figures used are incorrect for many combinations of materials, so how can we specify the correct value?

The real problem goes much deeper, in that there is actually no value that can be taken from somebody else's work and used for your own parts. This is the reason that this book has repeatedly suggested that all parts must be subjected to static test before you can be certain of their strength, regardless of the details of the construction.

Pursuing further the problem of exact strength, we can use some of the existing literature to illustrate the point. The publication ***Handbook of Composites***, edited by George Lubin and published by Van Nostrand Reinhold in 1982 (a new edition will be out in 1994 or 1995), has some strength data on E-glass fibers in various resin systems and at several fiber orientations and different resin content values. If one takes this information and reorganizes it into some simple graphs, we can see that stating a value for the strength of a simple composite structure becomes quite an exercise in guessing. Please bear in mind when considering the information shown in the following figures that strength figures for cross-plied laminates will usually be somewhat less than half the values that would be developed for a completely unidirectional layup.

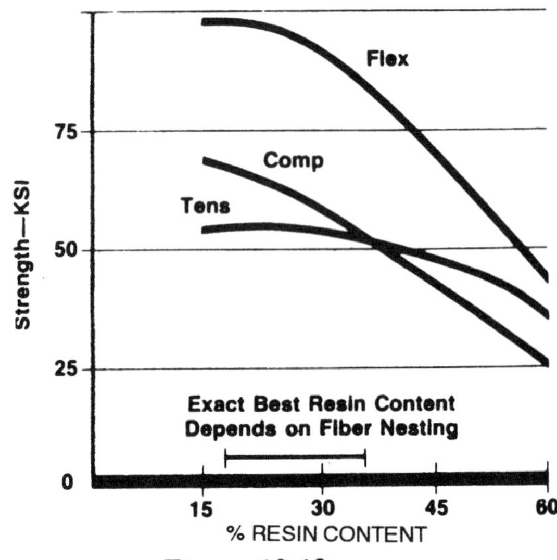

Figure 10-13.

Looking at Figure 10-13, the strength values for three different loading modes of a cross-plied laminate are plotted as a function of resin content (expressed as a percentage that the resin in the laminate weighs, as compared to the total weight of the laminate). The note along the bottom of the figure refers to the fact that even when using prepregs, some cloth weaves simply cannot be made to produce the lower resin-content laminates. For resin contents below about 35%, the only practical way to get there is to use a couple of layers of preimpregnated, collimated tape at right angles to each other, such as is done on Boeing floor panels, which routinely achieve these **low** ranges. The builder is usually stuck with a hand layup rather than prepregs, so you must look at the right end of the curves, as real parts usually fall in the range of 50% to 70% resin, judging from most of the parts I have seen.

Having uncovered our first can of worms, let's look at some of the other variables. Figure 10-14 shows the effect of changing resin systems, and is the basis for the standard recommendation that nobody should use the common polyester resins. They are at the extreme low end of the strength range, and only suitable for very low cost products such as bath tubs, shower stalls and small boats. A much better choice is to use either a vinyl ester or an epoxy for your matrix resin system.

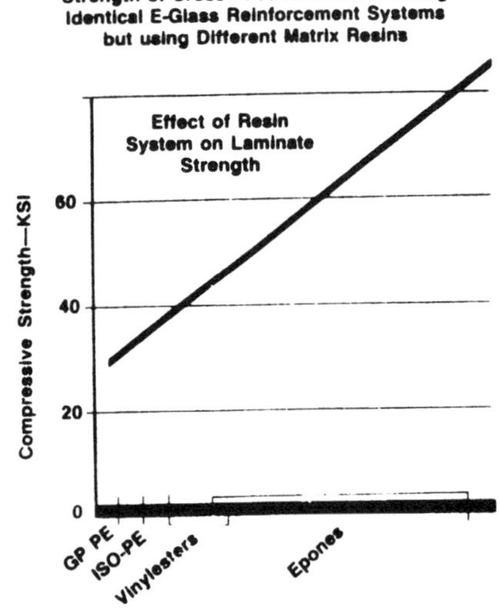

Figure 10-14.

Note that there is a wide range shown for the epoxy resins. The problem here is that the top end of this range is really not open to the wet laminate user because all the high-strength, high-toughness and high-durability systems require a high-temperature cure of at least 225° F, along with a cure pressure of 50 psi or higher, in order to achieve these published properties. The best systems for the home workplace are either the vinyl esters or the room-temperature curing epoxies. Fortunately, both of these systems will give **high** reliability and long life, even though they cannot deliver the high strength to weight ratios that we all would like to achieve.

The vinyl esters all show lower strength levels than most epoxies, but don't let that fool you. They have such a low viscosity and quick wet-out that you can usually get a little lower resin content with these than you can with the epoxies. As a result, the strength may turn out to be about the same, or even higher! This leaves us at the 35,000 to 40,000 psi level as the ultimate strength of our laminate, and envying the big guys who can afford prepregs and autoclaves to get those high values. There are still more variables to look at, however, and perhaps some of them can help.

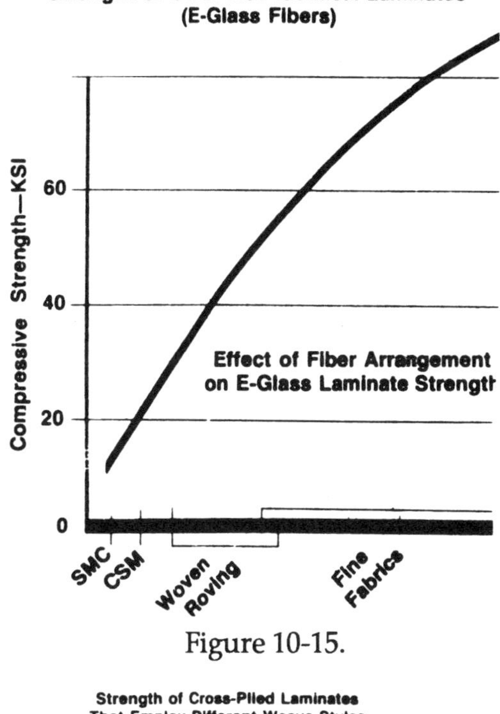

Figure 10-15.

Figure 10-15 shows the effect of changing the arrangement of the fibers in the reinforcing web. At the low end of the chart is the common Sheet Molding Compound (SMC) used by the truck industry for fenders, hoods and cabs on most of the big highway tractors. Next to it is the Chopped Strand Mat (CSM) which is used on most of the low-cost fishing, skiing or sailing boats. The really attractive strength levels don't show up until we get to woven roving and fine fabrics. These are all woven fabric plies, with the weight of the fabric nearly always a prime factor in the cost of the structure. These very heavy fabrics make the building of the structure go much faster, and therefore dominate the structures used in the better boats, in which higher strength and better quality is needed than that which can be obtained from the use of chop.

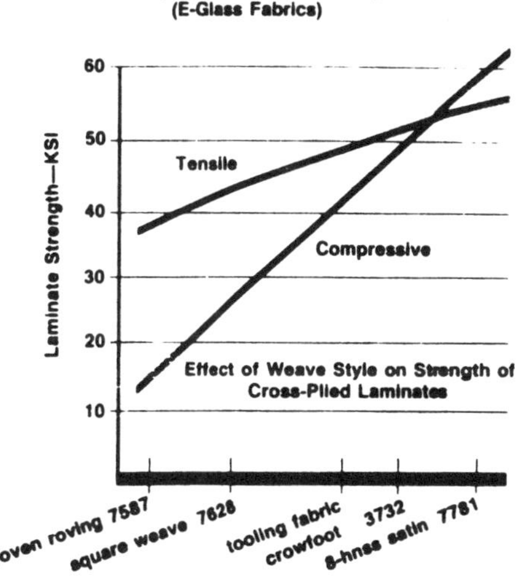

Figure 10-16.

Many homebuilt aircraft projects, and most boat projects, use fabric weaves that are at the lower end of the "fine fabric" range. Such fabrics as 602, 1002, style 7715 "UNI" and 7725 "BID" are common. About the only exception is style 7781 fabric, one of the best of the fine weaves, which is used throughout the structure of the Glasair and many other aircraft, including all Boeing designs.

The strength levels which will be developed by these weaves are not nearly so definite as the curves in published data would indicate. However, enough is known so that we can draw the curves on Figure 10-16 with the different weave styles in approximately the right relationship to each other.

Even though the information shown on the four curves, (Figure 10-13 through Figure 10-16) is quite revealing, and is believed by most authorities to be generally correct, it should be noted that much of the information used to draw them came from different sources, who used different techniques to arrive at their values, and usually did not know that another lab was working on a similar problem.

This multiplicity of sources gives both a better overall sense of credibility, along with a much lower degree of accuracy than might otherwise be expected from the data. Consequently, none of the charts has been dignified with an accurate scale and they should **not** be used for obtaining design data, but only for comparative purposes. There are several useful conclusions that can be drawn, however, even from such sketchy and unreliable information. Some of these conclusions are:

1. The same thickness of laminate, using the same fiber reinforcement, may yield strength values anywhere from as low as 20,000 psi up to as high as 80,000 psi, (or even higher), for a cross-plied laminate, depending upon just what resin system is used, what the resin content is, what form the reinforcement takes and, of course, whether there are any air pockets, voids, floor sweepings or other such trash molded into the part.

2. Less resin is always better, except when you have already reached the theoretical minimum resin content for the style of fabric you are using. This optimum resin content can be as low as 25% or so for unidirectional fiber layups, and as high as 50% or more for the open-weave fabrics like the boat and tooling fabrics, (usually called out as "6-ounce" or "10-ounce"). Most fine-weave fabrics that are designed as reinforcement weaves for prepregs, such as 7781, will have an optimum resin content of around 35%, although some of the square weaves may be much higher.

3. None of the curves allow for the effect of poor resin mixing, air entrapment in the laminate, or other such simple mistakes. **All** such mistakes make the laminate weaker, sometimes **much** weaker.

All of the charts refer to "cross-plied laminates," and mean that the layup was composed of half the fibers going at 0 degrees, and the other half going at 90 degrees to a reference line in the plane of the part. Many layups are made at other orientations, and the values shown in these curves would not apply to them.

In the case of unidirectional laminates, or those in which **all** of the fibers go in the same direction, such as the typical spar cap structure, the values will usually be about twice those shown on the charts.

Using this factor of two, it can be seen that the spar cap material made of E-glass unidirectional fibers could have a compressive strength of anywhere from a minimum value of about 30,000 psi up to a maximum value of perhaps 160,000 psi.

The more conservative designers stay near the low end of this range, since a one-off builder is not expected to wring out the high values that a Boeing or a Lockheed routinely demand.

Working backwards from some of the aircraft plans which have been sold, you can quickly calculate that one of the more reliable designers of composites for homebuilders does indeed use about 48,000 psi tension and 41,000 psi compression as design figures. The problem is that these figures are only valid if the **entire** system is used, so that you will really get a structure close to what the designer used. If you change the resin, the cloth weave, or the prescribed fabrication process, all bets are off, and your values will come in at a different figure, maybe lower, maybe higher.

HOW A BACK-YARD OPERATOR CAN GET REALLY HIGH ALLOWABLES

An inexperienced designer might become rather discouraged after reading all of the problems pointed out in the previous paragraphs. There is one method, however, by which anyone can obtain the startlingly high stress allowables enjoyed by the Boeings of our world, but without requiring all of the Process Control and Production Equipment normally required.

This path is the use of plies or rods made of material manufactured in a factory where all of the required methods are in place, except that the product is a simple ply of fiber in the form of a unidirectional flat ply (such as those made by Gordon Plastics or DFI Pultruded Products). These materials are produced by mixing the resin into the fibers while the fibers are under tension and being drawn into a closed and heated die in which the laminate is cured before leaving the die. This equipment forces out all of the unrequired resin, leaving only some very straight fibers buried in an absolute minimum of resin. Both manufacturers make their materials in both glass fiber and carbon fiber, and the structural performance is at the absolute top of the known range of values for these laminates.

Spar-Tuf (also known as "Bow-Tuf") has been used for many years as spar cap material, and Graphlite was developed a few years ago for use in the spar caps of the Bell-Boeing VS-22 "Osprey" tilt rotor aircraft. Neptco, who has withdrawn from the aircraft materials business, published the curve shown in Fig. 10-17, in which they attribute the high strength to the absence of waviness in the fiber. Although no corroborating data is available, the thrust of the data appears to be correct. Gordon Plastics publishes no data, but their E-glass material has been checked in the Hexcel laboratory, and was found to develop both tensile and compressive strengths higher than found in any other comparable material. (These figures were used for the very high strengths shown in Fig. 10-7 for glass laminate.) To make the situation even better, it is quite probable that the material will occupy only the very center of the "Normal Distribution Curve" shown in Fig. 10-18 for Material A. These are undoubtedly the reasons why a firm with a Process Control Department larger than most companies would choose a precured laminate made by a small supplier, instead of making their own laminates in the usual manner. Happily, there is no reason why a builder of sail boats, small aircraft or race car frames cannot do exactly the same thing!

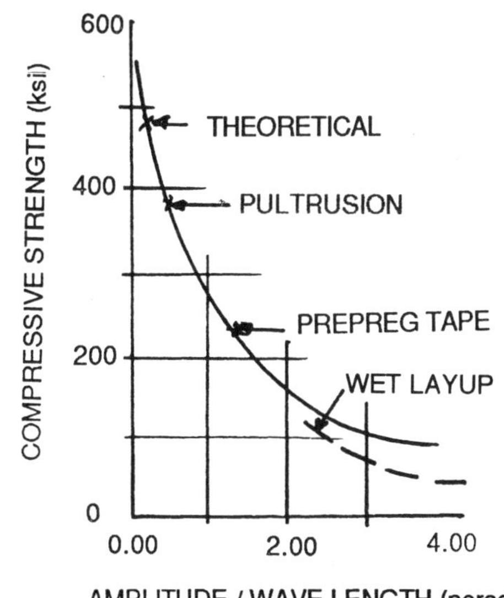

Figure 10-17.

VARIATIONS IN VALUES

In addition to the variations which result from different proportions and deployment of the materials involved, we should also look at just how consistently our structures will deliver a certain strength when we try to hold all the variables at a single value.

When we load a material in a testing machine, the load-indicating display of the machine quickly gives us a reading of exactly what the load was when the specimen failed.

If all the procedures are followed correctly and we are using a good machine, the value is usually accurate to within less than 1%, and we feel confident that we really do know what the strength of that particular piece was. When we put a second identical piece in the machine and load it to failure, however, the value we read may be considerably different than the value read on the machine for the first specimen. Even if there are 10 or 15 nominally identical specimens, this little surprise may be repeated for **every** specimen. Are we then to believe that the material is just no good, and cannot be trusted as something from which to make a structure? If you believe that to be the case, you will **never** find a material you can use. In the real world, literally all materials have this problem, and the only difference is one of degree.

Figure 10-18 is an entirely hypothetical (I made it all up) plot of the results of extensive testing of two different materials, each of which we might find in the average one-off project. The curve has been plotted in a special way, so that the horizontal scale is a strength scale, chosen so that a value of 100,000 psi falls in the center of the page.

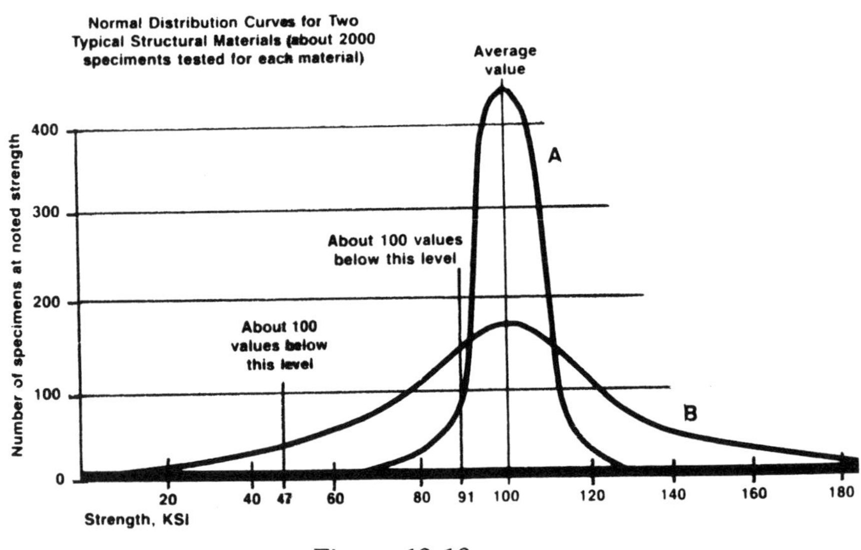

Figure 10-18.

The vertical scale is simply a linear scale showing the number of specimens that were found to have a strength value equal to one of the numbers on the horizontal scale. For simplicity, an average value of 100 ksi has been chosen for each of the two materials tested, and a "Normal Distribution Curve" has been plotted for each, based on tests of about 2000 specimens for each of the materials. To simplify matters, it is also assumed that our hypothetical lab technician running the test program has plotted as a single value all of the values that fall in a 4,000 pound range of values. Thus the point at the apex of the curve for material A shows that about 420 specimens had a test value of between 98,000 psi and 102,000 psi, and so on for the other points.

Material A has a rather tight spread between the lowest and the highest values found, about 80,000 to 125,000 psi. Remember however, that we have tested only about 2000 specimens and it may not be realistic to think that every piece of all material tested forever in the future will exhibit a strength above the level of the lowest one found in this test program. An even nastier reason that we should not require all material to fall above that 80 ksi number is that theory tells us that the curve does not have a definable lower end, and that we'd better decide on just what percentage of lower-than-desirable parts we can stand.

This firm fact derives from a science called "statistical analysis," which is now an integral part of every airframe manufacturer's Quality Assurance department. The math involved in statistical analysis permits the calculation of the various points of interest that might be found by plotting these curves, but lets you do it without the labor of making all those manual plots. The key point usually asked for, however, is the "standard deviation" of the plot, a point on the curve above which about 95% of all the specimens ever to be tested will fail.

When a critical application is involved, two standard deviations may be used, which results in about 97% of all the material being stronger, or even **three** standard deviations, which assures that about 99% of all the material will be stronger. Most material suppliers, however, do not give out data from which these figures can be calculated, so we are left to guess. Or if we are Boeing, we keep careful score, and after a few years we plot our own curves for every critical material in use.

For the builder of a single project, such a careful recording and calculation is just not possible, much less practical, as the structural part in question is being produced by the builder himself, and he will likely make only the single part now under construction.

If much production could be tested, however, most composite parts made by homebuilders would have a normal distribution curve more like that of Material B, rather than A, as shown in Figure 10-17. This is caused by unfamiliarity with the process and the materials, by variations resulting from the hand lay-up process, by too much or too little resin in the part, by air entrapment, by inadequate mixing of the resin, by erratic measurement of the resin components, and probably by many other things that have not been mentioned.

The end result is that we must be ruthless in using a lower value for design than that which we hope our part will actually demonstrate. The only way out of the problem is to plan to proof-load all critical parts, and simply not depend upon any pre-set or assumed value of strength at all. For starters, however, plan to use a preliminary value for strength of your parts that is **much** lower than the numbers which might be suggested by the curves in this book, or by any other source other than your own test results.

Before we can test our structure to some specific load, we must decide just what that test load should be. Let's look at some of the background and try to simplify the situation.

The term "limit load" arose from the fact that the body of aircraft structural knowledge in the 1930's had progressed to the point where most of the loads expected to be encountered by an aircraft in normal service were fairly well understood, and they were included in the then-new design requirements. The limit load was that load expected to be encountered in normal service, and had to be routinely carried without any structural deformation or impairment of subsequent operations of the various systems in the airplane.

The "ultimate load" was set at 1.5 times the limit load, partially because the yield point of aluminum alloys then in use tended to be about two-thirds of the ultimate strength, or conversely, the ultimate strength was about 1.5 times the yield strength. Thus, if a part could take a given load without permanent set, or bending and staying bent, it would take about half again as much to make it fail completely.

It was also good sense to make the breaking point of the structure well above the level expected in normal service. The original figures in the Federal Air Regulations have been revised many times as more information and experience has been developed in the areas of fatigue loads from low-level gusts, actual strength needed at the point of failure, degradation during normal service life, real requirements of special-use aircraft, such as the aerobatic designs, and so forth.

There has not, as of 1994, been substantial change as yet to make allowances for the all-composite aircraft. However, much thought is presently being given to the problem, and the all-composite Beech Star Ship has been certified and is in normal commercial operation. Also, both Boeing and Airbus have adopted composites for major parts of their newer commercial aircraft, and each is leading the certification authorities along a path which will eventually allow normal treatment of composite structures.

One of the principal problems is that the composite materials do not exhibit a yield point, followed by substantial deformation and a failure at a significantly higher load. Most composites, with the notable exception of Kevlar, exhibit a capability of carrying load within the elastic range, right up to the ultimate strength of the member, and then failing in an almost explosive manner. The designers who are familiar with this phenomenon show a healthy respect for the limitations of this type of material, and make allowances for its lack of yielding and deformation under **high** loads.

What is suggested in the face of all these problems is to use a single load factor to replace the limit load, ultimate load and safety factor. That is the reason for suggesting a load factor of 10 G's for homebuilt structures used in the example problems. Because of the nature of the way composites take loads, it is perfectly feasible to proof-load your structure up to 70%, or even 90%, of the ultimate load expected at failure, which is really the only way to be certain that the structure will take the loads it will be given in operation.

To be safe without this step means making a far heavier structure than might actually be needed to do the job. By using this step, you can avoid the use of a "safety factor" of two, at the possible expense of having to make another spar if yours should fail under proof load.

If you plan to fly your design without proof loading, even an extra factor of two might not pick up some of the variations in strength that are easily possible when you look at the wide spread in the values of strength properties shown by Material B in Figure 10-17.

In summary, the easiest way to think of the strength of a structure is simply as a multiple of the 1G load. The FAA says we need to have at least 3.8 G's for the Normal Category, or 6 G's for an aerobatic airplane, but let's just say that we would prefer to have the structure fail at something more than 10 G's when you give it a proof loading — or perhaps 12 G's if you plan to invite me along.

The point that is being repeatedly stressed is that each of the materials which one must choose as a designer has its own set of properties, its own advantages and its drawbacks, and that each material must be examined as a separate problem which might offer its own solution to make your design better.

The same admonition — TEST EVERYTHING — always applies.

CHAPTER 11

ADHESIVES AND BONDING

The art of gluing structures together is an old one, dating back several thousand years. Even today some of the old adhesives such as starch, animal hide glues and blood glues are still used extensively for some bonding jobs. The state of the art changed rapidly early in the last century when the science of resin chemistry began to attract enough chemists to make a dent in the overall problem of adhesion. The same breakthroughs that made advanced composites possible are the ones that brought modern adhesive technology to its present, highly usable form.

Until the 1940s, few people seriously considered using adhesives in the structures of modern aircraft except for wood, where adhesive bonding had for many years been the standard in the aircraft industry. Bonding metals is a much different problem than bonding wood structures, and even the adhesive bonds in wood did not always last a suitably long lifetime.

By the late 1930s, however, the resin chemists had not only discovered the casein and resorcinol adhesives that made high-strength bonded wood assemblies which lasted almost forever, but they had also found that the phenolics, when combined with either rubbers, or vinyl or butyl plastics, made good metal-to-metal adhesives. As a result, World War II saw many production-bonded wood structures that were quite reliable, as well as a few experimental bonded metal parts for aircraft.

The real beginning of today's bonded metal aircraft structures industry began after the war, when the post-war designs began to make use of the recent advances in resin chemistry.

By the mid-1950's, nearly every new military or large commercial aircraft employed adhesive bonding in significant areas of structure; by the early 1960s, the pressure to reduce structural weight had moved adhesives into primary structures on many aircraft. Helicopter manufacturers led the trend, due both to their greater need to save weight and to the remarkable improvement in fatigue strength that is nearly always achieved when bonding is substituted for other means of joining. This improvement in fatigue life led to successful all metal helicopter blades having both long service life and light weight. The leader at that time was Bell Helicopters, with the all-bonded tail rotor blade on their Model 47, and later in the 1960s, with a whole family of main rotor blade designs.

In large commercial aircraft, the key motivation was a near-disaster during the initial full-power ground engine test run of the B-52. This design originally had no bonded structure, but when the first three minutes of afterburner operation left most of the wing trailing edge structure on the ground in small, oval-shaped pieces of skin material, it became obvious that rivets had severe limitations under high sonic loading. With its "Dash Eighty" already in the works, the first B707 used a considerable amount of bonded structure even in the original version. Once this aircraft was operational, every subsequent airplane also used bonded structure in the same and other areas in order to compete at the same empty weight. The race was on!

Few homebuilders contemplate a project like a 707, however, so we will talk about some of the more familiar bonding problems.

Homebuilders still use many wood-based structures in both propellers and airframes, and many of the adhesive problems are much the same on these structures as they are on the more sophisticated metal airframes.

BONDED JOINTS IN WOOD

These joints were in common use before the Wright Brothers, but much has changed in the materials that are now available. Before discussing materials, let's consider a few of the important elements of a bonded joint. They are:

1. the joint design,
2. the fit and sequence of assembly,
3. the strength needed in the joint,
4. the subsequent environmental exposure of the joint,
5. the adhesive material to be used, and
6. the process to be used to complete the bond.

All of these elements must be addressed in a coordinated fashion, as each may have a bearing upon what decision is made regarding one or more of the other points. Most of the older literature on glued wood joints reflects the limited strength of the older adhesive materials used in the 1920s and 1930s. We are well advised to follow these practices now, even though the adhesives may be much better, since the use of the old practices achieves the best possible performance of the resultant bonded wood structure. The newer and stronger adhesives give us a more reliable, consistent and longer-lived joint. To do a poorer design job on the joint using stronger adhesives would result in an unknown reduction in strength, reliability, or both.

Element 2 in our above list is also little changed by the arrival of the new materials. It can be affected by the choice of adhesives, however, as some adhesives need a closer fit (poor gap-filling properties in the adhesive), and some adhesive materials give little time to the builder before they begin setting up. The allowable maximum "open assembly time" of your adhesive may well dictate a change in the assembly sequence or method if it is only a few minutes and you have a complicated joint to put together.

Elements 3 and 4 have changed quite a bit in the past 20 years. The Phenol-Resorcinol adhesives are stronger than nearly all woods, yielding shear strengths of up to 6000 psi or more, when tested with Maple or Birch blocks — a figure higher than those achievable with most epoxy formulations! Synthetic materials offer the joint a long life. Both the Phenol-Resorcinols and the epoxies offer freedom from degradation after long exposure to high humidity, fungus, bacteria, vermin, water soaking and, for some formulations, temperatures of 300° F and higher.

The adhesive material will also dictate, more or less, the process you must use. Some of the phenol-resorcinols and the epoxies need an elevated-temperature cure cycle, and in these materials many formulations are available that will perform well, after a room-temperature cure only. Some will accept either type of cure and yield almost the same properties. Since few homebuilders have equipment for hot curing bond lines, there is little interest among homebuilders in such materials. One of the interesting things about some of the newer formulations, however, is their ability to be cured at room temperature, and then exhibit excellent performance at service temperatures of more than 300° F. ("Newer" only applies to the epoxies because the phenol resorcinols have been doing this for more than fifty years!)

The other big difference in adhesive materials is the amount of pressure needed on the bond line in order to achieve good structural performance of the finished assembly. This can vary from no pressure, as in the case with most epoxies, to minimum pressures of 100 psi or higher, which is usually required for any of the phenolics.

Pressure may be needed for more than one reason. In the case of phenolic-containing formulations, the cure mechanism involves the creation of a water molecule as the reaction progresses, and if the cure is being carried out at 350 degrees F, a common cure temperature for these systems, the water becomes steam, generating about 100 psi of pressure. If the clamping forces are not higher than 100 psi, the result is often a blister or delamination in bond line, leaving an unusable part.

The second reason for using substantial pressure in gluing operations is that most of the parts to be glued together will not fit perfectly. This misfit, which nearly always exists to some extent, even on carefully prepared and fitted parts, results in a variation in thickness of the glue line. This variation can be tolerated by some adhesive formulations but not by others, even though they may be similar formulations of the same family of adhesives.

Some good adhesives become much weaker if the line thickness is more than a few thousandths of an inch. Thus, because many of the best wood adhesives exhibit this trait, it has become a fact of life for most experienced woodworkers to apply high pressure on the layup to ensure a minimum thickness bond line.

This ever-present requirement for high pressure in the bonded joint during cure can sometimes have some unfortunate side effects. For example, a low-viscosity, unfilled epoxy adhesive can flow out of the glue line when a high bonding pressure is used. This is usually true of most systems that are intended to be used as matrix resins in composites, as they are formulated to flow freely and wet out all the fibers without any pressure. When these materials are put between two hard, perfectly finished and perfectly fitted hardwood surfaces, and a high pressure is applied, they simply go somewhere else, leaving almost no adhesive and a very weak bond.

The solution of course, is to either use low pressure to clamp the joint, or add a filler material to the resin, such as Q-Cells, wood flour or glass beads, so that no amount of clamping pressure will make the adherend surfaces completely close together. A light fabric like cotton cheesecloth can also be used to obtain this "bond line control" feature in an otherwise unusable resin. (Be sure to wash out the starch-oil binder in the cheesecloth before using it!) The best system of course, is to use a formulation that is already set up for use with the clamping pressures you need for fit-up reasons, so if you produce a perfectly fitted joint, you will still get the final strength you wanted.

ADHESIVES FOR BONDING WOOD

POLYVINYL ACETATE EMULSION ADHESIVES - Typically, these are the "white glues" sold in a plastic squeeze bottle dispenser for household use, with several variants containing wood flour or ground nutshells and called "carpenter's glue." The version seen most often is Elmer's Glue made by Borden, one of the biggest suppliers in this family. These materials are low-cost and readily available, but suffer from being affected by moisture and heat. They should not be used for true structural applications. They are so easy to use that Americans bought about 333 **million** pounds in 1984, and the usage keeps rising.

CYANOACRYLATE ADHESIVES - these are the "one drop holds up an elephant" adhesives originally marketed by the Tennessee Eastman Company as its Eastman 910 series, and are now sold by many manufacturers under various names. Hysol has an entire new line of these materials that it calls "Superdrop" adhesives. They are characterized by nearly instant curing when applied to the area that is to be glued and then tightly clamped with modest pressure. They are wonderful for instant assembly (of nearly anything, not only wood), because they set so quickly that they can be held in place by hand while curing. The strengths achieved are not high enough to class them as structural adhesives, but for assembling bits and pieces of wood for use as a core in a complex sandwich assembly, they are quite satisfactory. These adhesives are sold in small containers of only 5 or 10 grams, and are **expensive** when compared to typical structural adhesives.

ANAEROBIC ADHESIVES - these materials are quite close in handling and performance to the Cyanoacrylates noted above, but are chemically different. The Anaerobics were pioneered by the Locktite Company, and are used for a variety of assembly operations in small metal parts, thread locking, nut and washer locking and other formerly unusual applications. Again, they are expensive compared to more conventional materials, but are used in rather tiny amounts in a typical application. The manufacturer claims that these materials are structural adhesives, but in the terms of the types of structures used to build airframe, they are not.

CASEIN GLUES - this is one of the most widely used wood glues because of its ease of storage, simple mixing with water and high structural strength. Some of the problems in its use are because it is a natural product derived from skim milk, and is not entirely free from loss of strength after exposure to moisture. These deficiencies are not enough to prevent its use in the primary structures of aircraft, but have resulted in its becoming a second or third choice for most such applications. It seems that most homebuilders would rather use the somewhat higher cost systems and take advantage of their better permanence.

PHENOL-RESORCINOL GLUES - Sometimes called resorcinol resin glues, these materials have been used since about 1943, and have almost entirely replaced the caseins and the ureas in the production of waterproof and marine-grade wood products. The typical product available to the homebuilder comes in a two-part container, perhaps labeled "Pennacolite," a trade name of the Koppers Company, it is available in better builders' supply and hardware stores.

The "A" part is a liquid resin of low viscosity and a dark brown or purple color; the "B" part is a light brown powder with a terrible phenol smell. The powder is carefully weighed for mixing with an accurately determined volume of the liquid, and the mixing is just as critical as it is with the more forgiving epoxies.

The amount of time available for completing the glue spreading and getting the parts into the assembly, as well as the time available for getting the assembly under pressure is limited, so only small amounts of the adhesive are usually mixed in a single batch. As with the casein adhesives, bond line control is usually provided by the ground walnut or pecan shells that are a part of the adhesive system. This allows you to simply apply an excess of adhesive to the joint and use a heavy clamping pressure to force out any unnecessary glue. The use of light pressure in this sort of adhesive system almost always results in much lower strengths in the joint, so in actual production practice, the clamping pressure is only limited by the compressive strength of the wood.

Although these adhesives are more demanding in their mixing requirements than the caseins, and cost more, their superior performance in comparison to most wood adhesives makes them the first choice of most professionals.

ADHESIVES FOR USE IN BOTH WOOD AND METAL

EPOXY JOINTS IN WOOD

The epoxies are not usually considered wood adhesives, although many formulations work quite well in this application. At least two problems are present when bonding wood with epoxies. The first is that of flow in the joint after application and clamping. Because most wood glues are formulated with a filler to provide control of bond line thickness, or are inherently too viscous to be forced completely out of the bond area under high clamping pressures, people tend to assume all bonding of wood joints must be done at high pressure. With the straight resin-and-hardener, (no fillers), present in many epoxy formulations, clamping tightly will force all of the bonding material to leave the joint, producing a low-strength bond.

Ironically, the problem is even worse when the fit-up is perfect and the wood surfaces are perfectly smooth. If the same thing were done on freshly sawed, unsanded and unplaned surfaces that have a lot of still-attached wood fibers hanging from the cut surface, these rough fibers would act like a bond line control additive, resulting in a much stronger joint. If there is any doubt, a scrim cloth of open weave glass or cotton may be used in the bond line to provide this control. Of course, the fiber used must be prepared in the same way as if it were to be used as a structural fiber in a composite. Washing is sufficient for cotton, but glass must have a surface finish compatible and appropriate to the resin system used.

The other problem with using epoxy to bond wood is the fact that many of the formulations are sensitive to moisture prior to reaching full cure, and most wood has a lot of moisture, even in the dry condition in which it is normally used for aircraft structures. As a result, it is not a good idea to assume that the epoxy you are using will be satisfactory as a wood adhesive. Check with the data sheet or ask the system's manufacturer before you make any bonded wood structures with it.

One notable exception to the problem of using epoxies as wood adhesives is the "West System" of adhesives and processes sold by the Gougeon Brothers Company in Bay City, Michigan. The company has worked its way through the problems of using these adhesives in wood construction, and those who have used its materials have nothing but praise for Gougeon Brothers. Remember though, when using these materials, to follow the process instructions, or you may not get the outstanding results reported by others.

EPOXY AND OTHER ADHESIVE JOINTS IN ALUMINUM

THE SURFACE PREPARATION PROBLEM

Using adhesives to produce structural joints in metal is different from accomplishing the same job in wood adherends. Woods offer a porous surface, while metals offer a completely non porous surface. Most wood adhesives make poor metal adhesives. The problem is metal's impervious and non-reactive surface, and the solution is to find adhesives that make an inherently high strength attachment to such a surface. This solution solves only part of the problem, as proper surface preparation of the metal must be considered. In this context, surface preparation means not only the removal of oils, grease, oxides and foreign materials, but the chemical adjustment of the metal surface so that the bond will be both high strength and long lasting. This requirement for chemical compatibility carries with it the implied requirement of a different chemical treatment for each combination of adhesive and adherend.

Some combinations are better understood than others, some give better permanent bonds and some surface treatments are suitable for several different adhesives, but in no case is the treatment identical for a different adherend. The adherend that has had the most extensive development effort is aluminum, as it makes up the majority of most aircraft structures built in the last 60 years. Since it is also one of the most desirable of the materials for making homebuilts, we will spend a few minutes discussing it.

The surface of a clean aluminum sheet has several things on it that usually go unnoticed by the casual observer, even assuming that any oil or grease has been removed. Under the grease film, however, is a layer of oxide of aluminum. This oxide may be either a hard surface such as that produced by an anodic treatment, or it may be a rather soft material such as that which forms naturally in air after the surface has been etched or cut. It is assumed that the soft oxide contains a water molecule and is probably not quite the same chemically as the hard oxide. The soft oxide may also exist in more than one form. At any rate, experts agree that it does not form a strong structure, and will not carry the same loads that can be carried by the parent metal.

It has been suggested that the weak oxide surface might be bypassed by anodizing the clean metal surface in a Chromic or Sulfuric acid anodizing process, thereby ensuring that none of the weak material would form on the surface. This does indeed work, and the subsequent formation of the weak oxides is permanently prevented. However, the anodized surfaces are rather difficult to bond to, and even with the best adhesives, do not result in the high strengths possible with some of the better surface preparation methods, even though you get a relatively permanent bond that rarely deteriorates with age.

With the passage of time, bonded aluminum joints made to perfectly clean but untreated aluminum surfaces experience a reduction in strength that is insidious. The adhesive shows an attractive and consistently high strength level immediately after cure, (whether room temperature or elevated temperature system), but then begins to show progressively lower strengths after longer exposure to normal storage conditions.

There is no clear relationship between the amount of time that has passed and the amount of reduction in the strength of the joint. It has been determined, however, that either the presence of high humidity or a high stress level will drastically accelerate the deterioration; if both are present, the loss of strength occurs even more rapidly.

This relationship was first clearly understood in the late 1960s when several airframe companies experienced rather alarming results in the very materials that were part of the primary structures of their aircraft. The work progressed simultaneously in many locations, but Bell Helicopter and Boeing-Seattle appear to be the companies that came up with a test so simple that it took about five years more than it should have to convince the other bonding companies that it was reliable.

This so-called "wedge test," shown in Figure 11-1 is now used in virtually every bonding shop in the entire airframe industry. It consists of a 1/8 inch thick sheet metal of the same aluminum alloy being used in the bonded structures, and cleaning it in the same metal preparation line used in production. A couple of 1 inch wide strips of this cleaned and prepared material are bonded together, using the adhesive and process at hand. After the bond is cured, the laminated strip, about 14 inches long, is placed in a chamber that has a spray-mist nozzle, which keeps the humidity above 95%, an electric heating element, (a light bulb works!), and a thermostat to hold the temperature inside the chamber at a constant 125 degrees F. The specimen is left in this environment for only one hour. Before it is put in, a wedge of the same 1/8 inch thick aluminum is driven into the bond line at one end of the strip, which forcibly splits the bond apart. The end of the split bond is marked with a scratch mark. After this assembly of specimen and brutally inserted wedge are removed from the "environmental exposure chamber" the specimen is examined to see how far the crack has progressed during the one-hour exposure.

If the distance between the original scratch mark, which marked the end of the crack before exposure, and the new scratch mark, which marks the end of the crack after

THE WEDGE TEST

The wedge-test coupon shown at right is typical of the tens of thousands tested by Boeing in the development and maintenance of the PAA process. Developed in 1970, it closely correlates laboratory tests to actual in-service failures.

The curves at right compare wedge-test data derived from laboratory coupons as well as coupons taken from in-service delaminated panels.

NOTE: The improved bond durability provided by PAA allows far more stringent acceptability levels than can be obtained using the FPL etch process.

(Courtesy Boeing Technology Services.)

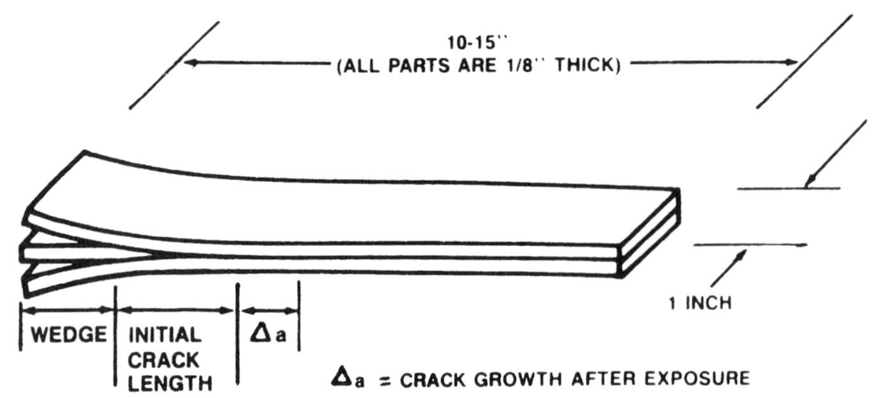

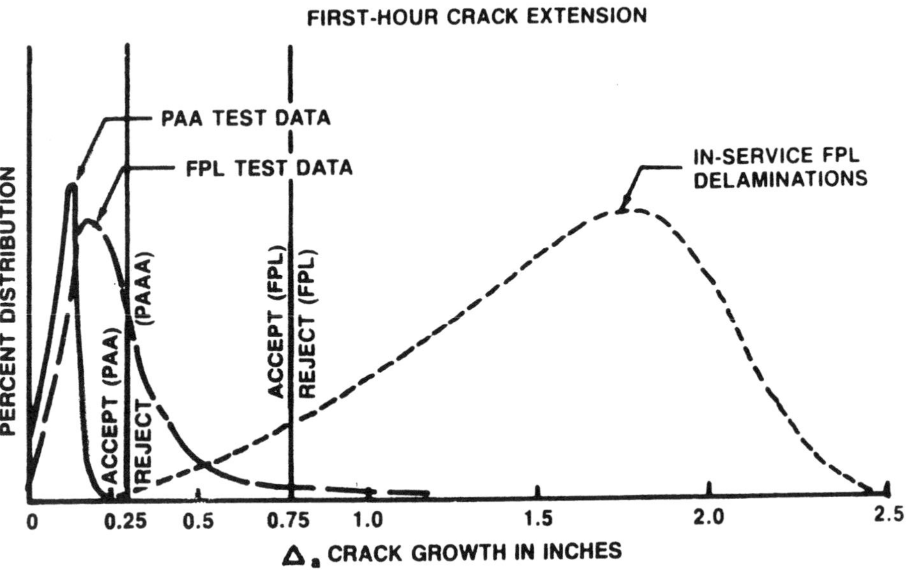

Figure 11-1. ASTM standard method 3762.

exposure, is less than half or three-quarters of an inch, the bonded joint is regarded as having a virtually infinite life in normal service. If the crack growth is approximately $1^1/_2$ inches or more, it is likely that the bond will show some service problems, and some delaminated parts are to be expected from this production batch.

If the bond was made without any chemical preparation of the surface, the strip will completely delaminate in the one-hour test! It is hard to believe that so simple a test has now replaced many millions of dollars of testing equipment in controlling the surface preparation of bonded aluminum parts, but this is now an accepted fact of life in the airframe industry.

We now know how to discern the good bonds from those that may fail, but how do we ensure that the parts we fabricate will last forever? In the case of aluminum bonding the only way is to adopt the same methods that the airframe makers use. Two processes are generally accepted and are broadly used at the present time: One is the "FPL etch," sometimes called a "sulfuric acid-dichromate etch," and the other is called Phosphoric

Acid Anodizing, or PAA. In either case, bond permanence is further ensured by applying a Corrosion Inhibiting Adhesive Primer, (CIAP), immediately after completion of the chemical surface preparation.

Using CIAP, also called CIP, is not only good practice, but removes the need to bond and cure the prepared surface within eight to sixteen hours after cleaning. With the use of either BR-127 primer, or the newer EA-9228 primer, (from American Cyanamid and Hysol, respectively), the cleaned and primed aluminum parts may be stored almost indefinitely prior to completing the bond, without any loss of subsequent bond durability.

By making use of a first-class surface preparation system with either of the above, you can then use aluminum inserts in your composite laminated structures almost without limit, and also without the likelihood of experiencing later delamination after aging. The major airframe builders such as Bell, Boeing and Raytheon, mix aluminum inserts as extra plies in their composite structures almost without limit. It is convenient to simply add an .020" layer of 2024 aluminum in a thin fiberglass edge, and use a few rivets where they might be needed for subsequent assembly.

Unfortunately, both the FPL etch and the PAA treatments involve extensive chemical control and a large investment in equipment. However, it is possible to find subcontractors in the bonding industry who will have an interest in working with either homebuilders or the kit manufacturers. This represents the only practical way to solve this problem.

The problem of surface preparation is actually much more complex than might be concluded from the simple situation described above. For a more complete review of the best current solutions to the specific problems of various adherends, the book, *Surface Preparation Techniques for Adhesive Bonding*, by Raymond F. Wegman, published by Noyes Publications, is strongly recommended.

CYANOACRYLATES

The same one-drop magic fluids that are used on wood also work on metals, but usually are not considered for a permanent, load-bearing structure because of their high cost, compared to the more commonly used materials, as well as their limited resistance to humidity.

ANAEROBIC ADHESIVES

These are primarily used for thread-locking, assembly of small complex metal assemblies and are not usually considered as structural adhesives for airframe parts. As is the case with the Cyanoacrylates, the cost is rather high, and they are used where they perform a unique job that cannot be done with one of the lower priced materials.

RUBBER-PHENOLICS AND VINYL-PHENOLICS

These are a huge family of elevated-temperature cure, (usually 350 degrees F), adhesives, usually furnished as a B-staged, supported film of non-tacky material, and still used today by the airframe makers, even though the modified epoxies are more common. When these materials are used in aircraft, they are required to meet Specification MIL-A-132 for metal-to-metal applications, and they must meet specification MIL-A-25463 if they are used in sandwich applications. As a matter of convenience, most of the B-staged film adhesives used by the airframe makers meet both of these specifications.

EPOXY-PHENOLICS, POLYIMIDES AND NITRILE RUBBER-PHENOLICS

These are three of the many families of adhesives used for elevated temperature applications in which the part will be used at 250 to 500°F. Some of the newer modified epoxies also have good strength at elevated service temperatures. Many of these materials are both expensive and difficult to use, and none of them, like the modified Phenolics listed in the previous group, are well suited for use by homebuilders.

ROOM TEMPERATURE CURE EPOXIES for ELEVATED TEMPERATURE SERVICE

This is a new class of adhesives and matrix resin systems that are practical for the homebuilder to handle in his own shop. The elevated temperature strength and toughness depend upon whose formulation you use, and whether you can give the finished part a post-cure at higher temperatures. The post-cure simply raises the bonded part to a higher temperature, usually in 50°F increments, up to about 350°F, after which the part yields a joint that exhibits high strength up to about 450°F. Since this post curing is done without tooling, only a hot-air oven is needed. Hysol EA934 is a typical adhesive, and Ren RP-4002 is a typical matrix resin of this type.

ROOM TEMPERATURE CURE EPOXIES

These are the materials that homebuilders use most often, since there are many manufacturers, ranging from 3M and CyTec Fiberitc, to the many smaller formulators. The materials are usually offered as a two-part system, and must be carefully weighed and mixed like all RT-cure epoxy systems. The differences between the various competing materials include the variations in viscosity, comparative resistance to various environments (such as gasoline), heat distortion temperature and the typical strength developed on a given adherend. Viscosities vary from a thin liquid, like H.B. Fuller's 2427 (what you often get when you ask for the obsolete Safe-T-Poxy, and which is a good adhesive), to the paste materials that are putty-like in consistency.

The single most preferred material among aircraft homebuilders seems to be Hysol EA 9410, which is a high-viscosity material that has both high strength and excellent tolerance for a wide variety of cure pressures, (down to zero psi is acceptable), bond line thickness and surface preparations. Hysol EA 9410 also meets the strength requirements of MIL-A-132 on several adherends. These materials tend to decrease in strength as the temperature increases, and at temperature levels of 150 to 200°F they have little strength. With this sort of material you should consult the data sheets in detail before making a decision of which one to use on your project.

ONE NOTE OF CAUTION: Do not use as structural adhesives any of the "hardware store" epoxies, the ones typically sold over the counter at retail stores and packaged in small squeeze tubes or double syringes. Some of them are very good materials, but little attention is paid by the manufacturer to what he puts into this week's production run, and no attention is paid to whether the strength properties at final cure meet any particular minimum level. In spite of this caution, I am sure that "5-Minute Epoxy" will be used in at least a few applications because it is just so convenient and available. But **do not** use it to glue your wing attach fittings on!

BONDING WITH OTHER ADHERENDS AND ADHESIVES

Mentioned earlier in this chapter was the fact that surface preparation is different for each material to be bonded. As might be judged from the length of this chapter and the fact that we have covered only two adherends and a few adhesives, a simple treatment of the subject cannot cover many combinations of materials in any usable detail.

A note of caution for those contemplating the use of one of the usual adhesives for bonding unusual adherends: **The strength of an adhesive joint varies widely for different adherends, even when using one of the excellent epoxies**, as shown in Table 11-1. The two most annoying cases are bonding to cured polyester laminates and to polycarbonates, like Lexan. The joint strength is almost always disappointingly low, and sometimes so poor that it is impossible to mount even simple shelves and brackets. If you cannot find the information for using your adhesive on the adherend at hand, make a few test bonds first to check the strength of your proposed joint.

TABLE 11-1

The following test values were obtained by Hysol, using the "Disc Shear Test Method, on the strength of their EA 9410, as reported in the Hysol Data Sheet on that adhesive.

SUBSTRATE	VALUE, psi	FAILURE
Aluminum	3580	Cohesive
Bonderized Steel	3780	Cohesive
Brass	4280	Cohesive
Cold Rolled Steel	3920	Cohesive
Copper	3690	Cohesive
Galvanized Steel	3790	Cohesive
Lexan	2020	Adhesive
Magnesium	3780	Cohesive
Nickel	3700	Cohesive
Nylon	880	Adhesive
Oak	1900	Substrate/Adhesive
Phenolic Laminate	3070	Substrate
Plexiglas	370	Adhesive
Polyester Laminate	3100	Substrate
Rigid PVC	360	Adhesive
Stainless Steel	940	Cohesive

NOTE:
1. All surfaces were sanded and solvent wiped.
2. Cured 24 hours at 77 degrees F.
3. Method was Disc Shear to Aluminum substrate.

For those interested in more factual material on this subject, several good references can be consulted, and it is strongly recommended that this be done.

A BRIEF GLOSSARY OF TERMS USED IN THE BONDING BUSINESS

GLUE - Any substance that sticks materials together with usable strength.

ADHESIVE - Glue that costs more than 50 cents per pound.

SYNTHETIC RESIN ADHESIVE - Glue that costs more than $5.00 per pound.

B-STAGED BONDING FILM - Glue furnished as a dry film, comes in a roll, and costs from $1.00 to $4.00 per square foot or more.

ADHEREND - The pieces that you glue together.

ADHESIVE FAILURE - The glue and the adherend separated cleanly leaving no residue on the adherend.

ANAEROBIC ADHESIVE - Expensive glue that cures almost instantly as a result of being confined within the bond line in the absence of air.

AUTOCLAVE - A pressure vessel into which the vacuum bagged part is placed so that the part is made to see not just the 15 or so psi that a vacuum provides, but up to as much as 500 psi or more.

BOND (noun) - The union of the materials you have glued together.

BOND (verb) - The act of gluing two things together.

BONDING FIXTURE - A device used to position exactly the adherends so that bushings, hinge lines, actuator points and other such items will be in the right place after the glue hardens. Also called a bonding tool or jig.

CLEAN SURFACE - Something *much* cleaner than you would expect.

CLEANUP - Almost impossible to do after cure if the glue is any good.

COHESIVE FAILURE - The glue failed within itself, leaving glue on each adherend.

CURE CYCLE - The specific sequence of temperatures, pressure and time used to cure a specific adhesive system.

DETERIORATION - Something bad that happens to a bond line.

DISBONDS - A polite way of referring to bonded assemblies that came unglued.

FORMULATED ADHESIVE - Glue made up of resin and hardeners, plus an infinite number of miscellaneous items, like mummy dust, diatomaceous earth, clay, flocked fibers, aluminum powder, asbestos fibers, ground rubber, and a million other **very important materials** that remain secret.

FORMULATOR - A person who formulates the glue, deciding just which of the million available additives to use for a specific product.

FPL - Forest Products Laboratory, an agency of the U. S. Government located in Madison, Wisconsin, and one of the world's leading source for authoritative information on glue, adhesives and bonding.

GLUE LINE or BOND LINE - The cured adhesive that is left between the adherends after gluing.

HYDROCLAVE - An autoclave that uses water instead of air or nitrogen for the working fluid, so that when it blows up, only a few hundred square feet of the factory are destroyed instead of an entire city block. It is also used up to about 1000 psi or higher, where a gas would be extremely dangerous.

PARTING AGENT - see RELEASE AGENT.

RELEASE AGENT - A surface film of non-stick material deposited on a tool or part to prevent glue or resin from sticking. It usually also prevents you from sticking anything, like paint, to the part that comes out of that tool.

RELEASE FILM - A release agent furnished as a film, like cellophane or Tedlar, that doesn't transfer to the part.

RELEASE PLY - A fabric treated so as to easily tear off a bonded or molded assembly and leave a clean, bondable surface behind it.

RESIN CHEMIST - A chemist who speaks an entirely different language than anybody else, including other chemists.

SERVICE DISBOND - A failed bonded assembly that got discovered by the customer after the part was put into service.

SQUEEZE-OUT - The adhesive that squeezes out the sides of the joint when you clamp it for curing. If there isn't any, you probably have a poor bond.

STARVED LINE - Almost no cured adhesive is left between the adherends after gluing, causing poor strength or disbonding.

STRAIGHT RESIN ADHESIVE - Glue made of only resin and hardener, but you don't know which ones.

SURFACE GLOSS - The typical shiny surface on a cured plastic or composite part that prevents the glue from sticking to it.

SURFACE PREPARATION - The cleaning and chemical preparation of an adherend so that it makes a strong and durable connection to the glue.

TEAR PLY - see RELEASE PLY.

VACUUM BAG - A device that excludes the air from a bonded assembly, thereby using the weight of all the air above the assembly to provide pressure to the bond line.

CHAPTER 12

LOAD TESTING YOUR STRUCTURE

In the foregoing series of chapters on composite structures, one of the most repeated suggestions is to be sure to test your finished structure so that you know it will develop adequate strength before you fly it. The suggestion has been made so that a partially done analysis or an untested material and process combination can be made into a safe-to-fly structure.

Aircraft Homebuilders are usually without access to the more sophisticated analysis methods used by the airframe companies. However, in addition to making use of this excellent capability to completely analyze the structures they design, these sophisticated airframe makers use static testing and proof loading on virtually every important piece of structure that they build for the first time. The methods they use are often quite expensive, but the same result, although not quite so accurately measured, can usually be attained with relatively primitive equipment and methods. This chapter will outline some of the testing equipment and methods that require only a limited budget to accomplish.

SAND BAGS AND WEIGHTS

The traditional method of applying weights to achieve a large total load over a substantial area is to pile on bags filled with ordinary sand, so that the required amount of load is applied to the structure simply by the dead weight of the sand. If you are a well-built person in good shape, that weight can be an even 100 pounds for each sack, and it only takes 10 of them to make a 1000 pound load. Most people, however, fill the sacks only to a weight of 50 pounds, which is much easier on the back. It may make the counting and the logistics a little harder, but it generally tends to keep more sand inside of the sacks.

The real problem, of course, is finding a source for the several hundred bags that it will take to make up the total wing spar load at the number of "G's" you have decided upon as a static test value. The most original solution to the problem that has been heard of is that of a man who rented a whole truck load of ready-mix cement packaged in the small 50 pound sacks, and returned them to the store after the load test was completed. Five gallon water containers, like the plastic versions used for camping, are also useful but there is the same problem of getting enough of them to make the large total weights usually needed for the total load.

Plastic bags filled with water work fairly well, but you should find the heavy ones used by the nurserymen and sand and gravel companies, if you are to avoid the disaster of broken bags and a flooded workshop.

MEASUREMENT OF WEIGHT

It is easy to forget about the exact weight requirement when you are busy filling bags with sand. Have a bathroom scale handy so that each and every bag gets weighed and marked as it is filled. Estimating the weight leaves so much room for error that it makes the whole static test procedure worthless. The accuracy of an ordinary bathroom scale is always subject to question, but you will find that it is surprisingly close. Also, most scales have an adjustment that allows them to be set to an accurate reading, provided that an accurately weighed standard of 50 or 100 pounds is available to use as a check from time to time.

HYDRAULIC JACKS

Another device for measuring large forces or weights is an ordinary hydraulic jack, in which the line between the pump and the cylinder has been connected with a tee to a pressure gauge. To estimate the range of the pressure gauge you will need, measure the diameter of the main cylinder, calculate its area, (half the diameter squared, times 3.14) and divide the jack's nominal rating by the area calculated. The gauge needed for a 5 ton Jack with a 2 inch diameter main cylinder would thus need a maximum pressure reading of 1 x 1 x 3.14 divided into 10,000, or 3183 psi. Of course, you cannot buy a gauge with this as a maximum reading, so you will have to settle for a gauge with the highest number on its face being 2500, or 3000, or maybe even 5000 psi. Any of these would work well.

The other necessary number is what the reading on the gauge will mean in determining the actual load that the jack is supporting. Looking at the number just calculated, 3183 psi. This will be the exact gauge reading when the load on the jack is exactly 10,000 pounds. In other words, each 1 pound on the gauge will mean that the jack is loaded with 10,000/3183 pounds, or 3.14, and conversely, when the jack has 1 pound on it, the gauge will read 1/3.14, or 0.3183 psi. This is somewhat confusing, so to prevent making errors you should make a new face for the gauge, in which the present location of 31.8 psi is replaced with a mark and the number "100 pounds," the location of 159.1 psi is marked "500 pounds," and so on, up to the capacity of the gauge. When this new face is glued over the old one and the cover glass replaced, you will have a valuable piece of heavy-duty loading equipment.

NOTE: A hydraulic jack is not intended to be used as a load-measuring device, and when it is, you will notice that the large amount of friction in the cylinder results in large variations in the pressure-gauge reading. To minimize this problem, tap the outside of the cylinder lightly with a hammer as you get close to the load you are trying to achieve. If the variations are still fairly large, the measurements you get will be in error by at least that amount.

LEVERS

Our modified jack is of no use to us unless we can figure out how to take care of the other two parts of the problem. These are the distribution of the load, and provision for the reaction, or the firm fixing in place of our test part while we apply the load. Load distribution is often handled with a form of lever called a "whiffle tree." Those who are my age and were raised on a farm are probably familiar with them, but for the benefit of our city friends, Figure 12-1 shows a "single tree," or a whiffle tree with only one loading point.

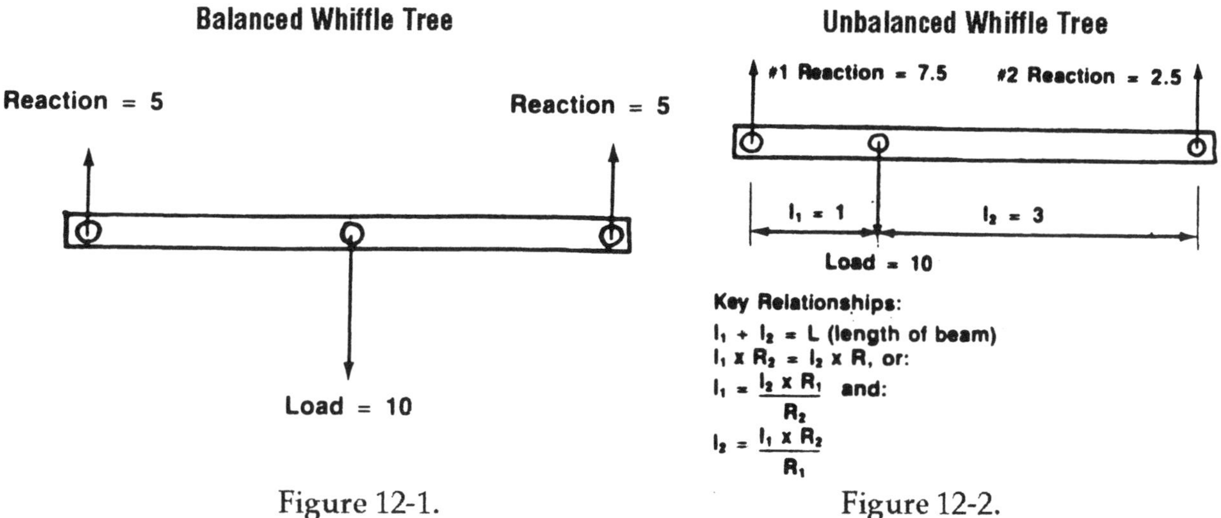

Figure 12-1. Figure 12-2.

As one can see, the single load point can be moved closer to one end or the other, and as it gets closer to one end point, the share of the total load taken by that point gets larger. The ratio of the two dimensions is used to calculate the exact number of pounds at each reaction point, or if we already have the reaction point loads, their ratio is used to set the distances so that the single load will exactly produce these two reactions, as shown in Figure 12-2.

When loading a wing or a tail surface, we like to use a distribution of the test load that is as close to the actual calculated load as is feasible. The usual compromise is to set the wing stations about a foot or two apart and calculate the total load to be carried by this local portion of the structure. This local total load is then applied to the spar, or to an area of the wing and spar near the center of the station used for that calculation. Remember that the location of the center of this load must be calculated not only for its spanwise location, but for the chordwise location, and that the test load actually applied must be centered reasonable close to this calculated location.

Many older references describe sandbag loading of wing spars as a structure separate from the rest of the wing structure. While this gives accurate results for some structures, it is usually better to test the entire wing structure, including the leading and trailing edge structure. These additions to the spar, especially in most composite designs, tend to change both the strength and the deflection of the spar, and although the change is often for the better, this is not always true. Also, in many composite designs, the resistance of the wing structure to torsional deflection is almost entirely a result of adding the outside composite skin to the spar and rib (or core) internal structure. The torsional deflection can be of primary importance, especially if a torsionally "soft" wing is combined with a spar location aft of the center of pressure, which causes the twist to magnify the effect of gust loads.

Another troublesome point is the effect of wing deflection on aileron and flap hinges and controls. If deflection of the wing under load is more than a few inches, or if you have unusual control features, it might be advisable to hook up the entire system, so that it can be checked for proper rig and performance when the wing is under limit load.

After the exact number of pounds and exact location of each load to be carried at each station is determined, the whiffle trees are designed to pick up each of these loads in pairs, and a secondary whiffle tree is designed to pick up each adjoining pair of whiffle trees, and a tertiary set of whiffle trees designed to pick up each pair of secondary whiffle trees, and so on, until finally all the load points are picked up by a single load equal to the total of all the point loads in all the load stations that were calculated. Where an odd number of stations exists, one of the secondary whiffle trees simply picks up a whiffle tree load on one end, and the primary load of a station point at the other end. Figure 12-3 shows such a set-up for testing an imaginary wing structure. In most cases, you will need only a three tiered set of whiffle trees to load the entire half-span of a wing, and fewer for a stabilizer or a fuselage.

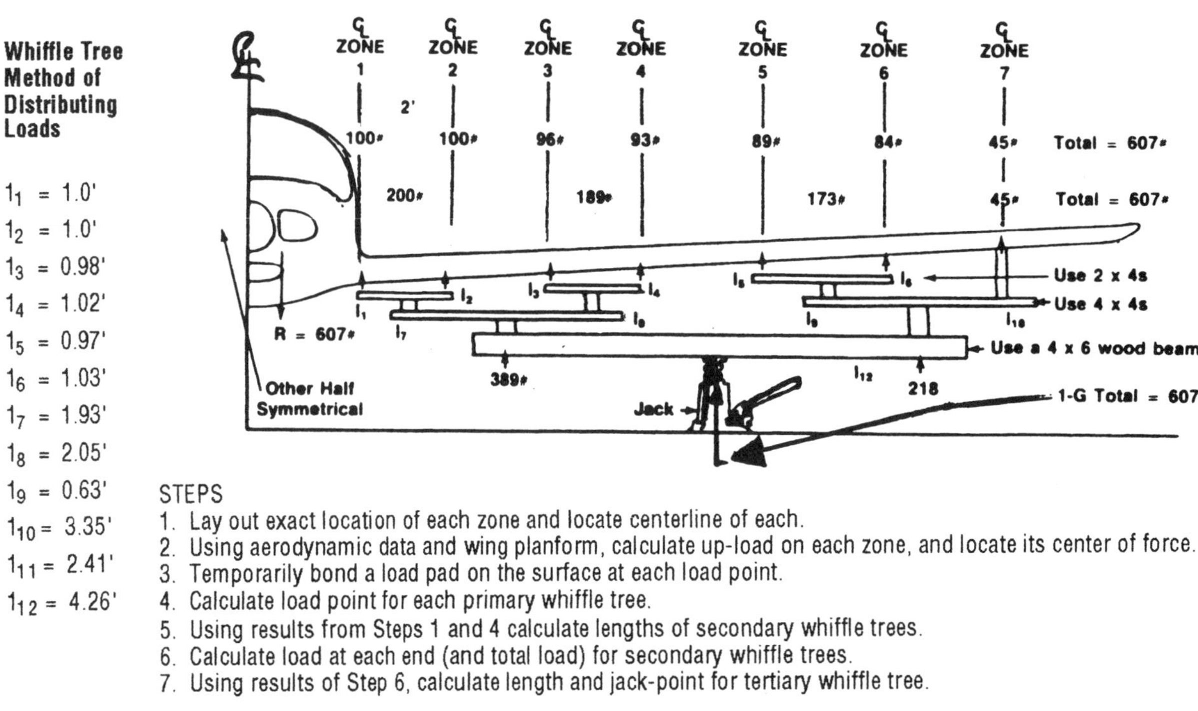

Figure 12-3.

WHERE THE LOAD ENTERS THE STRUCTURE

In the case of sandbags or water bags, the load enters the structure through the entire area in which the bag is in contact with the structure. Other than putting a blanket over the wing surface to keep it from being scratched, most composite wing or tail surfaces need no special protection. With whiffle trees, however, the load at a single point can be high enough to cause dents in the skin or local crushing of the core underneath. To prevent such problems, a pad, having sufficient area to distribute this point load over a reasonable area, must be used. Since you will want to complete static-load testing prior to your final finish work on the outside of the structure, these pads can be temporarily attached with a little bondo or micro. This will hold the pads in the location where they belong, permit the accurate assembly of a complex test set-up and allow subsequent removal before finishing.

This temporary attachment of the test points also permits you to choose between an upright or inverted position of the structure during the test. Where the surface is more than slightly contoured, make a load pad to fit the surface, as shown in Figure 12-4. Where the load-point is to be directly over a spar cap, a 4 inch square plywood pad is ample, but where the load point is away from this sort of "hardpoint," use at least a 6 or 8 inch square, or larger, if the structure has thin facings.

RIGHT-SIDE UP, OR UPSIDE DOWN?

When we prepare a structure for a static-load test, the part to be loaded can be positioned whichever way is the simplest, as long as the loads it sees are in the same relative direction that they will be in when the structure sees actual loads in service. Thus, if we decide on sandbags or cement sacks to provide the test load, the load can act only in the "down" direction, and the usual lifting force of the air is simulated by mounting the structure upside down and placing the sacks directly upon the now conveniently located lower airfoil surface. Most sketches in the literature show only a couple of saw horses supporting a wing or spar center section, but don't be misled. The load you will be supporting may be as high as 15,000 pounds or more, and if that saw horse gives way, it will result in a disaster! Plan to weld up a sturdy fixture that will be steady under at least **10 times** the total load planned.

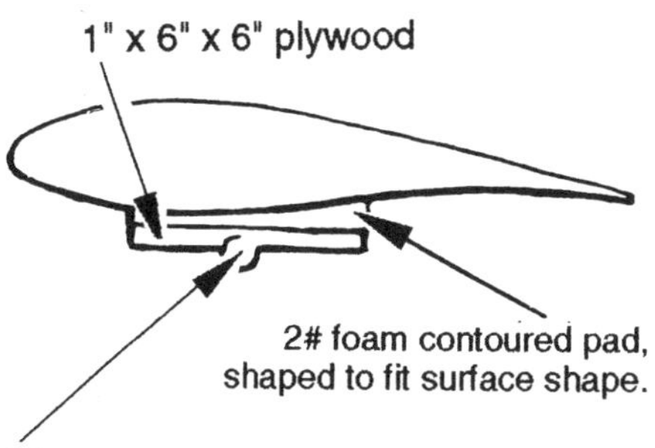

Figure 12-4

If you plan to use a jack-type load application, the structure can be tested in the right-side-up position, with the reaction point simply tied down to a suitably strong anchor in the floor. Again, **do not** underestimate the magnitude of the forces involved. Even those beefy-looking 3/4 inch anchor bolts at your local hardware store **cannot** be trusted to develop more than 2500 pounds or so. If you load a single point of your garage floor with much more weight, you'll pull out a piece of concrete.

Be particularly wary of concrete floors with no reinforcing bars in them. They are **fragile**! The benefits of this method is that when something gives, whether it be the anchor point or the structure under test, little secondary damage results.

With the sandbag method, a failure, once started, tends to accelerate, and makes repairing and retesting more work. If you use the sandbag method, it is prudent to provide ample support just below the structure under test so that a failure will merely result in the structure moving onto these supports, rather than crashing to the floor.

LEVERS

The use of levers need not be confined to whiffle trees. If a substantial wood beam, say a 4 x 6 about 12 feet long is used in conjunction with a set of four bathroom scales and a

A Method Of Using Low-capacity Scales To Weigh Large Loads

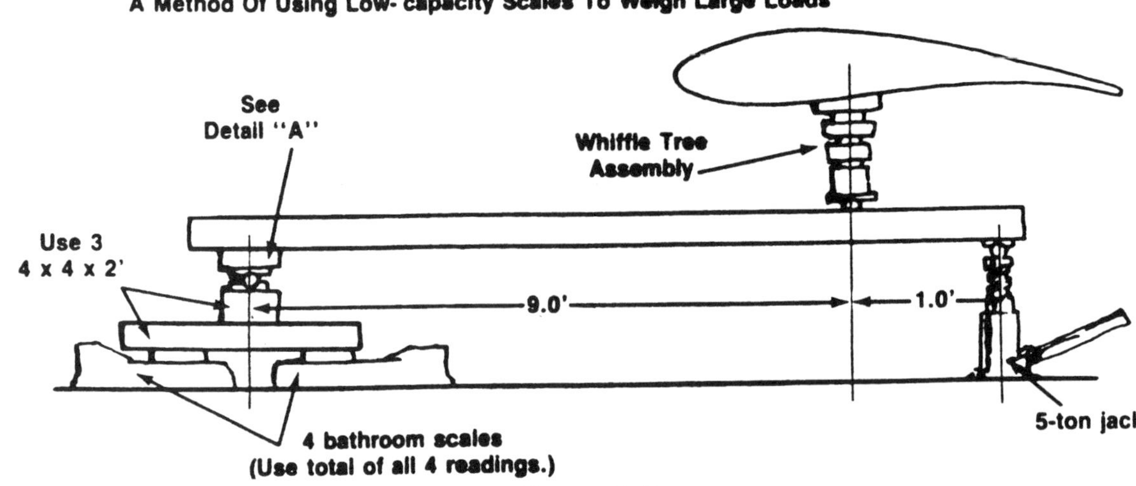

Figure 12-5.

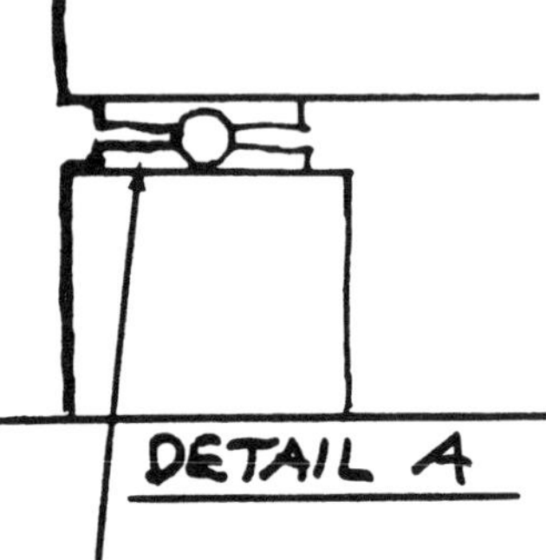

3" x 3" x 3/4" steel plates, drill 3/4" diameter x 3/16" deed to locate 3/4" diameter steel ball bearing point. Add soft foam between plates to prevent wobbling when not under load. Use at all three load points on main beam so as to accurately establish 9.0' and 1.0' dimensions.

jack, a test load of 10,000 pounds can be both easily generated and accurately measured.

If you measure the load on only one semi-span, and use a simple jack and support on the opposite side in order to keep the deflections and loads in balance, this would allow testing of the wing of a 2000 pound gross-weight airplane up to a simulated 10 G load!

For smaller craft and testing only to a limit load of 4.4 or 6.0, two bathroom scales would probably be sufficient. Figure 12-5 shows how such an arrangement might be laid out for the static test of a small aircraft wing. Figure 12-6 shows how a static test of a fuselage might be made.

Note that the total of the weight shown in Figure 12-6 is less than the gross weight of the aircraft. The weight of the outer wings, plus anything they may carry, is supported by the lift generated by the wings themselves, and does not pass through to the fuselage during flight.

REACTION PROBLEMS

In the above cases, remember that for every pound the hydraulic jack pushes on the floor, some other place on the floor must resist being pulled up. For loads up to 1000 to 2000 pounds, a simple expanding-type anchor bolt will work well. For loads upward of 2000 pounds, however, plan to use a spread-out pattern of anchor bolts, at least six feet apart. Too much pull-out load over too small an area will soon result in a requirement for a new garage floor. The ideal solution is to borrow a 12 or 14 foot length of a steel I beam of at least 8 inch depth, and attach both load and reaction points to this beam. If the length is more than half your wing span, it should be ample.

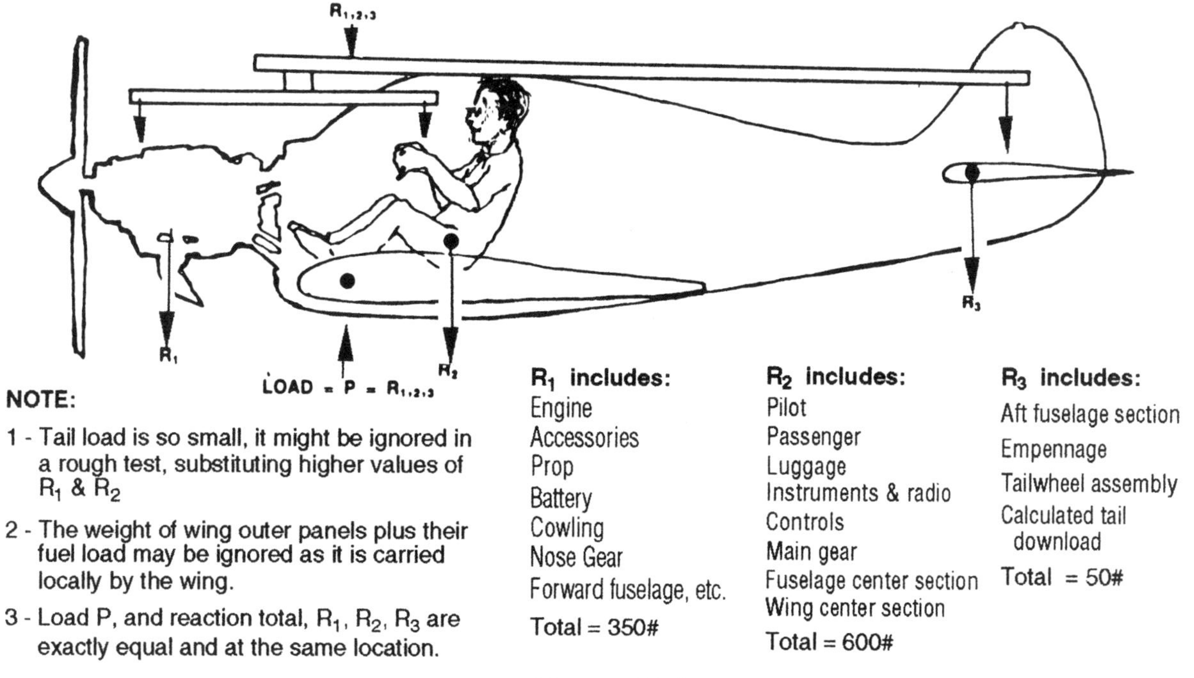

Figure 12-6.

TEST RESULTS

Your test program should demonstrate the operating or limit load that you expect your structure to routinely withstand. Past experience also tells that some additional load capacity beyond this figure is advisable, since we can expect some errors in our test results, as well as some possible future deterioration in the structure itself. For all-composite structures, or for composite/metal structures in which the metal parts are aluminum with a suitable surface preparation, this extra margin need only be about 10%. Test load for a utility category wing would then be 4.4 x 1.1, or 4.84 G's, or 6.6 G's for an aerobatic airplane.

If there is some question as to the materials, surface preparation, processes used, use of corrosion-inhibiting adhesive primer (or lack of same), or any other such question, **no suitable factor can be applied**. Where these questions exist, a static test should be repeated at every annual inspection. Only this re-testing can ensure that the strength level remains at a value that is appropriate to the use of the aircraft.
Deflections at key points such as the wingtip should always be measured and recorded. The first reason is to assure that no permanent set remains after the load is removed. If there is permanent set, the structure is regarded as having failed the test, even though no catastrophic failure has occurred. However, be sure to carefully examine and analyze the load and support or anchor systems. This is where the problem often comes from, rather than from a structural problem in the aircraft under test.

The second reason is that structural deterioration with advancing age is often visible as an **increase** in deflection at the same load. Even a 1G or 2G load will pick this up, and these lower loads are much easier to deal with as an annual inspection method.

MATERIALS TESTING

Materials testing of small pieces, rather than completed assemblies, can be very useful in choosing between materials, or determining if a similar material has a higher or a lower strength. The most commonly used such test for composites is the Flex Test.

Common practice in the glass-weaving industry does not include the structural testing of laminates made from the fabrics produced. Quality control is exercised by checking the strength, weight and uniformity of the yarn to be used before weaving begins, and by checking the yarn count, fabric area weight in ounces per square yard, tensile strength in both the warp and fill directions, and the uniformity and appearance of the final fabric produced. Subsequent checks are then made for specific requirements, such as the finish applied after heat-cleaning.

However, large volumes of material may be produced between such measurements. Thus, neither Hexcel nor Burlington will normally perform any structural testing of a laminate made from the fabric they are producing, but will only examine the fabric itself. As a consequence, the effect of using a different material in the manufacture of any composite made from the fabric must be determined by the maker of the composite. The weavers plead that all they should be expected to do is keep producing the same product **without any changes at all,** since each of their customers (end users) may have worked out a different resin system and manufacturing process, and are depending upon all of the fabric-related variables to remain constant.

When the user of the fabric is a Boeing or an IBM (printed circuit boards are one of the largest users of glass fabrics), it sets up its own specifications, which cover all of the receiving inspection tests, resin compatibility tests and laminate property tests that are felt to be necessary. A person operating in a home workshop does not have this capability.

Testing machines are rather costly, and only the larger firms own them. Even the least expensive machines cost several thousand dollars, and a typical budget figure for a small capacity machine might be between $10,000 and $100,000. The more expensive units will be much higher, depending upon the size of the machine and the degree of automation and sophistication desired. Some companies really get carried away! It is possible for a single machine and accessory package to run close to one million dollars, even at a capacity of only one hundred thousand pounds or so.

These expensive testing machines are not really necessary for the testing we will do, as long as we can accept errors on the order of 10% or so in the values we get. In order to achieve any sort of testing reliability, however, we must exercise careful control over the dimensions of the specimen we will manufacture, and will also need to make up a set of accurately machined test fixtures.

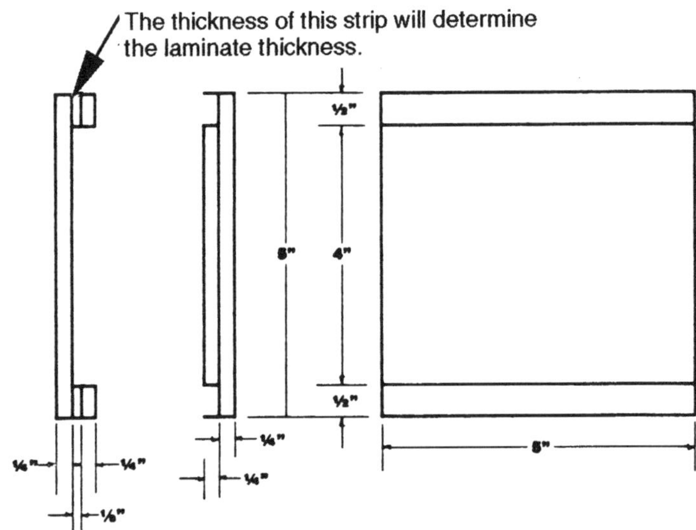

Figure 12-7. A laminating mold for making flex test specimens.

First, we must establish the piece of laminate we will test. The thickness of this laminate is one of the most sensitive variables we will control, and a variation in this thickness is a common cause of widely varying test values. For the test geometry selected in this case, our specimen must be **exactly** 0.125 inches thick. An error of only 0.005 inches in this dimension will give an **8% error** in the calculated flex strength.

To make such accuracy in the cured thickness practical, we will need a mold in which to lay up and cure the test piece, even though it is a simple, flat laminate. Figure 12-7 shows a design for a mold in which a 4 inch wide test piece can be laid up to yield several narrow strips for testing. Apply an effective mold release to the surfaces that come in contact with the resin, or you will have a terrible time removing the cured piece. If you don't have any Frecote 44 or one of the other standard mold releases, a layer of Saran Wrap will usually do nicely.

In using the mold, wet out and lay in place enough plies of your fabric to be as near to 0.125 inches for the total number of plies as the fabric thickness will allow. Use a little excess resin in the layup and let the pressure of the top plate squeeze it out, so that the layup will be exactly the same thickness as the gap strips on the mold. If you are in doubt as to how many plies will be required, put an extra ply in the layup, and measure the laminate thickness after cure to see if the layup prevented the top plate from closing down completely on the side strips. If the final thickness is more than 0.125 inches, use enough fewer plies so that the mold can close to the 0.125 inch thickness.

To save time and material, this experimenting can be done on dry plies of fabric before the resin is applied. Remember, if you use too many plies, the strength of the specimen will be higher, either because you have squeezed the layup down to get an artificially high glass content and low resin content, or because the mold did not completely close and you have an over-thickness specimen. In either case, the flex test results will be unusable, although the extra high numbers might make you feel good.

If we are to obtain strength values that reasonably represent the open mold, hand layup produced materials you are using to build an airframe, we must have the fabric layers under only modest pressure when the mold is closed. It should be rather difficult to close with your bare hands, but quite easily closed with modest pressure from a C-clamp or a vise. Fabric layers should be carefully cut to exactly fit the mold cavity, or be slightly undersized. The mold itself, shown in Figure 12-7, has open ends so that excess resin is forced out the ends of the layup as the mold is closed.

Once you have established how many plies of fabric are to be used to give the exact thickness of the mold cavity, always use the same number of plies for subsequent specimens of the same cloth style. This allows resin to make up for minor variations in fabric thickness. (There **will be** a little variation in the thickness of different fabrics of the same style, but at the most, this should be less than 0.001 inches per ply.)

When making a layup of unidirectional fabric, such as 7715, be sure to cut the width of each piece close to the mold cavity width. A width that is too narrow will allow the closing pressure to spread out the plies of fabric, and you will not really have a sample that represents the number of plies in the mold. If the strips are too wide, they will wrinkle up when squeezed into the narrower cavity, resulting in low and widely varied test results when you later load the specimens. Careful workmanship in making test samples has a direct effect on both the strength levels and the spread in the actual values measured, just as it does on your airplane structure.

After the test layup has cured, remove it from the mold and give it a post-cure of about four hours or so at 200 degrees F, (or better yet, use the manufacturers recommendations) in an ordinary cooking oven. Shield it from direct exposure to flame or electric elements. This step will permit a shorter test program period, and will tend to duplicate the strength of nonpostcured parts that have been sitting around for weeks or months. It will also give you a clear measure of any increase in strength that your resin system might yield as a result of the postcure.

When the molded laminate is ready, cut it into strips slightly wider than 1 inch, and sand or grind them down to exactly 1.00 inches wide. Remember that the fiber direction for a unidirectional laminate must be **exactly** parallel to the edge of the specimen. For a cross-plied laminate, the yarns in the fabric must be oriented exactly at zero degrees and 90 degrees to the specimen length. The cut edges must be finished with a belt sander, or other method that will leave a nice finish, since ragged or torn edges on the test parts will yield substantially lower values. Don't waste your time trimming or dressing the ends of the specimen, as they will not affect the results, even if they are a mess.

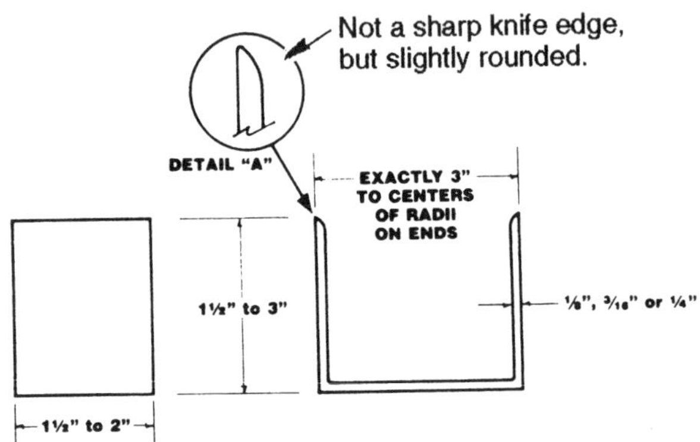

Figure 12-8. A load reaction fixture for testing flex specimens.

The next items of equipment we will need are the support and loading devices. Figure 12-8 shows the cradle that will support the beam as we load it. It can be welded up from 1/8 inch thick steel plate, or simply sawed off a length of 3-inch steel channel. In either case the **exact centers** of the radii on the upper ends of the supporting flanges must be precisely 3 inches apart, and exactly parallel with each other, which means that the flanges of a 3 inch channel must be bent slightly outwards so that the center of the radius of the rounded tips are made to reach the 3 inch dimension. Again, careful workmanship is a must if the fixture is to give usable results. Provide an inch or so

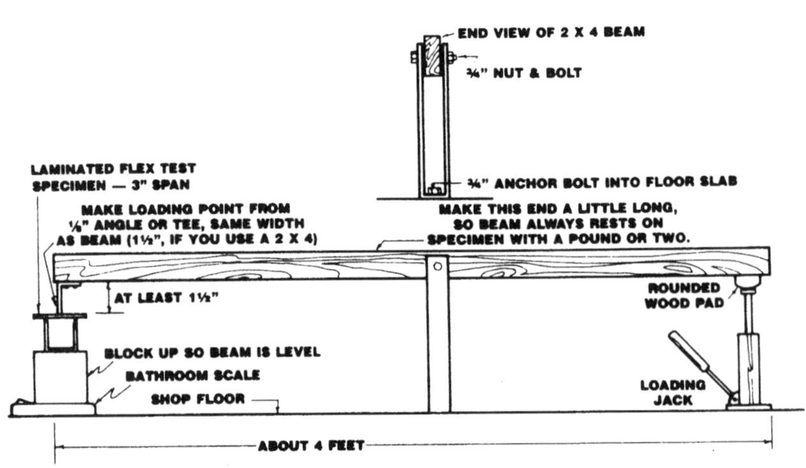

Figure 12-9. A home-built testing machine.

for the deflection at the center of the test specimen, as these specimens will bend a startling amount before breaking.

Figure 12-9 shows the loading beam. The pivot point in the center of the beam is not involved in our calculations, so it doesn't need to be anything fancy, except that it will be required to carry several hundred pounds when we apply the load to the specimen. Make the end with the loading point an inch or two longer than the jack end, so that it will be slightly heavy, and hold the specimen in position after you set it in place.

You now have the complete set-up, and are ready to make a flexure test. On one end, the test specimen is set up in its support frame on a bathroom scale, with some blocking to position it at the height that will make the beam level. On the other end is a small hydraulic jack, or any other means of obtaining the 200 or 300 pounds of upward force we will need.

Be sure to do a nice, neat and workmanlike job on all of the parts of the testing machine. All of the support and loading edges should be straight and parallel to each other and to the beam's pivot axis. The strap-iron bracket holding the beam pivot bolt should be rigid, to allow no movement of the pivot bolt in any direction as the load is applied. Make the beam long enough so that the nonvertical travel of the loading point is small. The 2 foot distance shown between the pivot and the loading point is the shortest that you should use, and will keep the lateral motion below 0.001 inch, for a 1/2 inch deflection of the test specimen.

The formula used in calculating the load on the surface fibers of the test specimen looks like this:

$$F = 3\,PS/2W \times t \times t$$

where:

F is the fiber stress in bending at the surface at failure (ultimate stress), in pounds per square inch, or psi;

P is the load, in pounds, that we will read on the scales;

S is the span, in inches, that the specimen is bridging (3 inches, in our case);

W is the specimen width, in inches; and

t is the specimen thickness, in inches.

For our specimen geometry then,

$$F = [P \times 3 \times 3] / [1 \times 2 \times .125 \times .125]$$

or:

$$F = 288\,P$$

As you can easily calculate, a stress in the specimen of 60,000 psi will mean that we will apply a load of 208 pounds. Since flexure specimens usually yield somewhat higher stress values than the compressive stress values we use in design, your load at failure may well exceed the capacity of the scales. If it does, simply cut down the width of the specimen to 1/2 inch instead of 1 inch.

The formula will then read:

$$F = [P \times 3 \times 3]/[2 \times 0.5 \times .125 \times .125]$$

or:

$$F = 576\ P$$

When we do this, the same 208 pound load will make the specimen see a stress on the surface of 120,000 psi, which is probably higher than you will achieve with a wet layup of unidirectional E glass. If you are testing some stronger material, reduce the specimen width. Be careful, however, as the narrower specimens are more sensitive to variations in the dimension and quality of finish on the edges. You will probably not obtain usable results at widths of less than 1/4 inch, and even at that width, there would be considerably more scatter in the test results. A better solution for stronger materials is to use two bathroom scales with a beam across them, with the load reaction fixture in the middle of the beam. Using four scales would also be better than making smaller specimens.

APPENDIX 1

SOURCES OF MATERIALS AND INFORMATION

One of the more perplexing problems facing the composite builder or designer is that of tracking down materials and information. Because the entire industry is relatively new, and the technology has only recently found its way into universities and public libraries, many honest attempts to find material and process information result in either a high degree of frustration, or no results at all. In Appendix 1 you will find some of the more productive paths and names of some specific sources. Addresses and telephone numbers for all of the organizations mentioned are listed in a separate section at the end of this Appendix.

THE SOURCES

Information you will need comes from several places. Because of the rapid growth and recent advances of this technology, much of the information is available from the material manufacturers, but not yet covered by the more traditional texts and references. Sometimes several manufacturers have similar information, but differ a bit on the presented facts, even when they are such simple things as describing the weight of a standard fabric. Thus, it is a good idea to obtain information from at least two sources for any important data. In addition to the manufacturers of a material, or a class of materials, distributors of the same products may also be a good source of information, and are listed separately.

Available reference books continue to be few in number. Several are listed in this Appendix, but keep an eye out for others. There are new ones being published every few months, some quite good.

One excellent family of information are the various specifications written by branches and agencies of the U.S.Government. Because of repeated budget tightening, many Government agencies which used to be a convenient source of MIL Specs, MIL Handbooks and other such publications are no longer easy to work with. However, an excellent new source has appeared. It is a privately owned firm named,"Document Engineering Company", or DECO. This firm now actually answers telephone calls with a real person, and will sell and ship you any of these publications on the basis of your Credit Card number! Their free telephone number is (800) 645-7732. If you are forgetful, it also dials correctly as (800) MIL-SPEC! In addition to all of the government publications, DECO also carries publications from many of our technical societies, such as SAE, ASTM, and even non-U.S documents, such as ISO, DIN and many others.

CORE MANUFACTURERS and DISTRIBUTORS

Many of the makers of core materials are not well known outside the industry they serve, even though some are rather large corporations. In addition, some cores, such as Nomex honeycomb, are made by only a few manufacturers, while others, such as Kraft Paper honeycomb or Urethane foams are made by many. Following is a list of core manufacturers, listed by the type of core they make.

ALUMINUM HONEYCOMB
- Alcore
- Hexcel
- Plascore
- Unicel

CARBON FIBER HONEYCOMB
- Hexcel
- Ultracor, Inc.

BALSA WOOD
- Alcore
- Baltek

GLASS FIBER REINFORCED PLASTIC HONEYCOMB
- Alcore
- Hexcel
- Ultracor, Inc.

KOREX ARAMID FIBER HONEYCOMB
- Hexcel

MILITARY-GRADE KRAFT PAPER HONEYCOMB
- Hexcel
- Unicel

NOMEX HONEYCOMB
- Alcore
- Euro Composites
- Hexcel
- M. C. Gill
- Plascore

PLASTIC TUBE CORE
- Alcore
- Plascore

ESTER FOAM
- Diab Corp.

POLYMETHACRYLIMIDE FOAM (Rohacel)
- Richmond Aircraft Products
- Rohm Chemische

POLYSTYRENE FOAM
- Aircraft Spruce & Specialty
- Dow Chemical

POLYURETHANE FOAM
- General Plastics

PVC FOAM

Baltek
DIAB Corp.

SAN FOAM

ATC Chemical

STAINLESS STEEL HONEYCOMB

Aeronca
Astech
Kentucky Metal Products
Rohr

SYNTACTIC FOAM

Bryte Technologies
Emerson & Cumming
Loctite Corp.
3M Company
YLA, Inc.

WEAVERS

There are many weavers of fine fabrics, woven roving and various unidirectional fabrics. Some of them will only respond to large requirements, but even these firms will direct you to a supply house that can handle small or sample quantities. A few of the weavers are listed below.

Cytec Fiberite Corp.
Hexcel Corp.
JPS Glass Fabrics
Textile Products, Inc.

SOURCES OF WET RESINS AND ADHESIVES

The formulators who provide resin systems for use in composite facings include both large, integrated chemical companies, as well as small firms that make no basic resins at all. No real distinction can be made between them, as some of the most advanced formulations sometimes come from those suppliers who make no basic resin materials themselves.

Ashland Chemical
Bryte Technologies
Composite One
Cytec Fiberite
Dow Chemical
Fiber Resin Corp.
Gougen Brothers
Loctite Corp.
3M Company
SIA Adhesives
Vantico, Inc. (formerly Ren Plastics)
YLA, Inc.

MANUFACTURERS of PRE-IMPREGNATED STRUCTURAL FABRICS and TAPES

ARC Technologies
Bryte Technologies
Cytec Fiberite
Hexcel
Loctite Corp
Newport Composites
YLA, Inc.

DISTRIBUTORS of FABRICS

Aircraft Spruce & Specialty Company
Airtech International
CMI/Composite Materials Inc.
Hawkeye Enterprises
National Aerospace Supply Co.
Richmond Aircraft Products
Wicks Aircraft Supply

DISTRIBUTORS of RESINS, ADHESIVES and PREPREGS

Aircraft Spruce & Specialty Company
Composites One
E. V. Roberts
Fiber-Resin Corp.
Gougen Bros.
National Aerospace Supply Co.

DISTRIBUTORS of NOMEX and ALUMINUM HONEYCOMB

CMI/Composite Materials Inc.
Texas Almet

SOURCES of PRE-CURED UNIDIRECTIONAL LAMINATES

DFI Pultruded Composites, Inc.
Gordon Plastics

DISTRIBUTORS of VACUUM BAG LAYUP SUPPLIES

Airtech International
Hawkeye Enterprises
National Aerospace Supply Co.
Richmond Aircraft Supply
Taconic

SOURCES for TOOLS that cut KEVLAR FABRICS and LAMINATES

Aircraft Spruce & Specialty Co.
Dayton-Price Ltd.
National Aerospace Supply Co.
Technology Associates

MAKERS of REINFORCING FIBERS for COMPOSITES

Glass Fibers

Owens-Corning Fiberglass Corp.
PPG Industries

Carbon and Graphite Fibers

Cytec Carbon Fibers
Grafil, Inc.
Hexcel
Nippon Graphite Fiber
Toray
Zoltek

Kevlar Fibers

Dupont

DESIGN and STRESS ANALYSIS

Aircraft Design
Antrim Associates

FUNDAMENTAL INFORMATION ON COMPOSITES

Composite Airframe Structures, by Michael C. Y. Niu
Technical Book Company, 2056 Westwood Blvd., Los Angeles, CA 90025

International Encyclopedia of Composites, Stuart M. Lee, Editor, VCH Publishers

Handbook of Composites, S. T. Peters, Editor, Chapman & Hall

ASSISTANCE ON MATERIALS & PROCESSING OF COMPOSITES

Composite Technology Consultants, Inc.
Wm. F. Condon, President
12310 Northwinder Row, Bayonet Point, FL 34667
727-863-5559

OTHER PUBLICATIONS of INTEREST

The AR-5 Tapes
The Arnold Company, 5960 S. Land Park Drive, #361, Sacramento, CA 95822

Light Airplane Design, by L. Pazmany
Pazmany Aircraft Corp., PO Box 80051S, San Diego, CA. 92138

A Practical Guide to Airplane Performance and Design, by Donald R. Crawford
Crawford Aviation, PO Box 1262, Torrance, CA 90505

Composite Aircraft Design, by Martin Hollmann, $14.95
Aircraft Designs, Inc., 5 Harris Court, Building 5, Monterey, CA 93940

Formulas For Stress and Strain
by R. J. Roark
McGraw Hill Books

Elements of Sport Airplane Design
by P. E. Bird
Vogel Aviation, PO Box 54, Reseda, CA. 91335

Low-Power Laminar Aircraft Design
by Alex Strovnik
2337 East Manhattan Drive, Tempe, AZ 85282

Fluid Dynamic Drag (Out of print, try your County Library)
by S.F. Hoerner

Fluid Dynamic Lift (Out of print, try your County Library)
by S.F. Hoerner

Fabricating Tips (regarding Derakane vinylester resins)
Ashland Chemical

Analysis & Design of Flight Vehicle Structures
by E. F. Bruhn
S. R. Jacobs & Associates, Inc., 9135 Madison Street, Suite B, Indianapolis, IN 46260

MIL-C-81986, Core Material, Plastic Honeycomb, Nylon Paper Base for Aircraft Structural Applications
DECO, 15210 Stagg Street, Van Nuys, CA 91405
(800) 645-7732

MIL-C-7438, Core Material, Aluminum for Sandwich Construction
(Address same as above)

MIL-C-8073, Core Material, Plastic Honeycomb, Laminated Glass Fabric Base, for Aircraft Structural and Electronic Applications
(Address same as above)

MIL-Y-1140, Yarn, Cord, Sleeving, Cloth and Tape-Glass
(Address same as above)

MIL-HDBK-17, Composite Materials Handbook
(Address same as above)

Mil-HDBK-23, Basic Methods for Stress Analysis of Sandwich Structures
(No longer in print--see your County Library; however, the information contained in MIL-HDBK-23 is about to be included in the new version of MIL-HDBK-17)

ARP-1524, Specification for Phosphoric Acid Anodize Surface Preparation Method
Society of Automotive Engineers or DECO

Epoxy Plastic Tooling Manual
Ren Plastics

Bulletin G-600, Preparation of Surfaces for Adhesive Bonding
Dexter Aerospace

TSB-120, Mechanical Properties of Hexcel Honeycomb Materials
Hexcel

TSB-124, The Basics on Bonded Sandwich Construction
Hexcel

Kevlar -Aramid Fiber For Light Aircraft Construction
Dupont

Korex Aramid Honeycomb Cores
Hexcel

PBJ , a computer program for calculating stresses in sandwich laminates, including data bases for most materials.
Antrim Associates

Handbook of Lay-Up and Bagging
by William W. McCracken
To order, e-mail <mj.mccracken@gte.net> or call 562-210-1429

COMPANY NAMES and ADDRESSES

Aerocell
12806 Scout Ave.
Bell Gardens CA 90201
562-927-2546

Aeronca, Inc.
1712 Germantown Road
Middletown OH 45042
513-422-2751

Aircraft Design
c/o Martin Hollman
5 Harris Ct., Bldg. 5
Monterey CA 93947
831-899-1510

Aircraft Spruce & Specialty (East)
900 South Pine Hill Road
Griffin GA 30223
800-831-2949

Aircraft Spruce & Specialty (West)
PO Box 4000
225 Airport Circle
Corona CA 91720
877-477-7823

Airtech International
5700 Skylab Road
Huntington Beach CA 92647
714-899-8100

Albany International Techniweave
112 Airport Drive
Rochester NH 03867
603-330-5837

Alcan Baltek Corp.
10 Fairway Ct.
Northvale NJ 07647
201-367-1161

Alcore, Inc.
1324 Brass Mill Road
Belcamp MD 21017
410-676-7100

Aldila Materials Technology
13450 Stowe Dr.
Poway CA 92064-6871
858-4 86-6970

American Composite Manufacturers Assn.
(ACMA)
1655 N. Fort Myer Drive, Ste. 510
Arlington VA 22209
703-525-0511

Antrim Associates
4018 Archery Way
El Sobrante, CA 94803
Tel: 510-223-9680

Bryte Technologies
18410 Butterfield Ave.
Morgan Hill, CA 95037
408-776-0700

Burnham Composites
4203 W. Harry
Wichita KS 67209
316-946-5900

Composites One
11917 Altamar Place
Santa Fe Springs, CA 90670
800-237-0087

Cytec Engineered Materials
208 E. Technology Circle, Ste. 300
Tempe AZ 84284
480-730-2311

Cytec Engineered Materials
1440 N. Kraemer Blvd.
Anaheim CA 92806
714-630-9400

DFI Pultruded Composites, Inc.
1600 Dolwick Drive
Erlanger, KY 41018
859-282-7300

DIAB Inc.
315 Sea Hawk Drive
Desoto, TX 75115
972-228-7600

Dow Chemical
2040 Dow Center
Midland, MI 48674
800-441-4369

Dupont, Composites Division
Highway 1
McBee SC 29101
800-951-3456

Emerson & Cumming
59 Walpole St.
Canton, MA 02021-1838
781-821-4250

Euro Composites, S.A.
13213 Airpark Drive
Elkwood, VA 22718
540-727-8502

E. V. Roberts
18027 Bishop Ave.
Carson CA 90746
310-204-6159

Fabric Development
1217 Mill St.
Quakertown PA 18951
215-536-1420

General Plastics Manufacturing
P.O. Box 9097
Tacoma, WA 98409
206-623-2795

Gordon Composites
2350 Air Park Way
Montrose CO 81401
800-399-0757

Gougeon Brothers, Inc.
100 Patterson Ave.
P.O. Box 908
Bay City MI 48707-0908
989-684-7286

H. B. Fuller
3530 North Lexington Avenue
Shoreview, MI 55126
800-328-7307

JPS Glass Fabrics
201 Slater Road
Slater, SC 29683
800-288-0577

M.C. Gill Corp.
4056 Easy Street
El Monte, CA 91731
626-443-4022

Gordon Composites
2350 Air Park Way
Montrose, CO 81401
800-399-0757

Grafil Inc.
5900 88th Street
Sacramento, CA 95828
916-386-1733

Hawkeye International
PO Box 451485
Los Angeles, CA 90045
310-338-5575

Hexcel Corp.
11711 Dublin Blvd.
Dublin, CA, 94568
925-847-9500

Huntsman Advanced Material
(formerly Ren Plastics)
4917 Dawn Ave.
E. Lansing, MI 48823
517-351-5900

Kozloff Enterprises
18021 J Skypark Circle, Suite 1012
Irvine, CA 92714
949-786-7742

Loctite Aerospace Division
PO Box 312
Baypoint CA 94565
925-458-8243

Marshall Consulting
720 Appaloosa Drive
Walnut Creek, CA 94596
925-945-6051

National Aerospace Supply Co.
33155 Camino Capistrano, Ste. C
San Juan CA 92675
949-240-6353

Owens Corning Corp.
Fiberglas Tower
Toledo, OH 43659
864-296-4054

Plascore
615 North Fairview Ave.
Zeeland, MI 49464
616-772-1220

PPG Industries
One PPG Place
Pittsburgh, PA 15272
412-434-3131

Richmond Aircraft Products
13503 Pumice Street
Norwalk, CA 90650
562-404-2440

Schnee Morehead
111 South Nursery Rd.
Irving TX 75060
800-878-7876 or 972-438-9111

Society of Automotive Engineers
400 Commonwealth Drive
Warrendale, PA. 15096
724-776-4841

Taconic
Process Materials Division
3070 Skyway Drive, Bldg. 203
Santa Maria, CA 93455
800-832-0982

Texas Almet
2800 E. Randol Mill Rd.
Arlington, TX 76011
817-640-2345

Textile Products, Inc.
2512 West Woodland Drive
Anaheim, CA 92801
714-761-0401

Toray Carbon Fibers Inc.
700 Parker Square, Ste. 205
Flower Mound TX 75028
972-899-2930

Ultracor
136 Wright Brothers Avenue
Livermore, CA 94550
925-454-3010

Unicel Corp.
1449 Simpson Way
Escondido, CA. 92025
760-741-3912

YLA, Inc.
2970 C Bay Vista Court
Benicia, CA 94510
707-747-2750

APPENDIX 2

GLASS WEAVE STYLES CROSS REFERENCE

Burlington Industries B.I.
Clark Schwebel C.S.
J. P. Stevens J.P.S.
Uniglass U.
Hexcel Corp. H.C.

Weave Style	B.I.	C.S.	J.P.S.	U.M.	H.C.	Yarn Count	Construction Warp	Construction Fill	Type of Weave	Weight oz/sq-yd	Thickness in. (appx)
32	3732	332	332	--	--	48 x 32	37 1/0	37 1/0	Crow	12.96	0.013
80	1671	1680	880	--	--	72 x 70	150 1/0	150 1/0	Satin	5.70	0.006
100	--	--	--	103	--	80 x 70	900 1/0	900 1/0	Plain	0.77	0.002
102	1042	325	325	--	--	88 x 44	900 1/0	900 1/0	Plain	0.74	0.001
103	100	100	100	--	--	80 x 70	900 1/0	900 1/0	Plain	0.77	0.002
104	--	--	--	--	--	62 x 52	900 1/0	1800 1/0	Plain	0.56	0.002
106	--	--	--	--	--	56 x 56	900 1/0	900 1/0	Plain	0.73	0.002
107	--	--	--	--	--	60 x 35	900 1/2	900 1/0	Plain	1.04	0.002
108	--	--	--	--	--	60 x 47	900 1/2	900 1/2	Plain	1.43	0.002
112	--	--	--	--	--	40 x 39	450 1/2	450 1/2	Plain	2.10	0.003
113	--	--	--	--	--	60 x 64	450 1/2	900 1/2	Plain	2.46	0.003
116	--	--	--	--	--	60 x 58	450 1/2	450 1/2	Plain	3.16	0.004
117	--	121	121	1912	--	54 x 39	450 1/2	450 1/2	Plain	2.49	0.003
118	--	--	--	--	--	90 x 60	450 1/2	450 1/2	Crow	4.02	0.005
119	--	--	--	--	--	54 x 50	450 1/2	450 1/2	Plain	2.79	0.004
120	--	--	--	--	--	60 x 58	450 1/2	450 1/2	Crow	3.16	0.004
121	--	--	--	1912	--	54 x 39	450 1/2	450 1/2	Plain	2.49	0.003
122	1522	1522	1522	--	--	24 x 22	150 1/2	150 1/2	Plain	3.69	0.006
125	--	--	--	--	--	36 x 34	450 2/2	450 2/2	Plain	3.75	0.005
126	--	--	--	--	--	34 x 32	450 3/2	450 3/2	Plain	5.45	0.007
127	--	--	--	--	--	42 x 32	450 3/2	450 3/2	Plain	5.94	0.007
128	--	--	--	--	--	42 x 32	225 1/3	225 1/3	Plain	5.94	0.007
132	--	--	--	--	--	48 x 32	150 2/2	150 2/2	Crow	13.00	0.015
135	--	--	--	--	--	42 x 33	75 1/0	75 1/0	Crow	6.00	0.007
138	--	--	--	--	--	64 x 60	450 2/2	450 2/2	Crow	6.64	0.007
139	--	--	1385	831	--	64 x 60	225 1/2	225 1/2	Crow	6.64	0.007
141	--	--	--	--	--	32 x 21	225 3/2	225 3/2	Plain	8.80	0.011
143	--	--	--	--	--	48 x 30	225 3/2	450 1/2	Crow	8.78	0.009
162	--	--	--	--	--	28 x 16	225 2/5	225 2/5	Plain	12.20	0.015
164	--	--	--	--	--	20 x 18	225 4/3	225 4/3	Plain	12.72	0.016
181	--	--	--	--	--	57 x 54	225 1/3	225 1/3	Satin	8.92	0.009
182	--	--	--	--	--	60 x 56	225 2/2	225 2/2	Satin	12.40	0.013
183	--	--	--	--	--	54 x 48	225 3/2	225 3/2	Satin	16.75	0.018
184	--	--	--	--	--	42 x 36	225 4/3	225 4/3	Satin	25.40	0.027
190	--	--	--	--	--	20 x 10	225 1/3	225 1/3	Leno	2.36	0.006
191	--	--	--	--	--	20 x 10	225 1/2	225 1/2	Leno	1.65	0.005
192	1610	1610	1610	--	--	32 x 28	150 1/0	150 1/0	Plain	2.48	0.004
194	1562	1562	1562	--	--	30 x 16	150 1/0	150 1/0	Leno	1.93	0.005
195	--	--	--	--	--	20 x 24	150 1/0	150 1/0	Leno	1.90	0.005
196	1620	1620	1620	--	--	20 x 20	150 1/0	150 1/0	Plain	1.60	0.004
196	1621	--	--	--	--	30 x 14	150 1/0	150 1/0	Leno	2.36	0.006
198	1658	1658	1658	--	--	20 x 10	150 1/0	75 1/0	Plain	1.60	0.004

Weave Style	B.I.	C.S.	J.P.S.	U.M.	H.C.	Yarn Count	Construction Warp	Construction Fill	Type of Weave	Weight oz/sq-yd	Thickness in. (appx)
202	--	--	--	--	--	20 x 10	150 1/0	150 1/0	Leno	1.20	0.004
203	1659	1659	1659	--	--	20 x 10	150 1/0	75 1/0	Leno	1.60	0.004
208	1668	--	--	--	--	60 x 10	150 1/0	225 1/0	Plain	2.86	0.004
209	1229	--	--	--	--	50 x 20	450 1/0	450 1/0	Plain	0.94	0.002
216	2185	--	2116	--	--	60 x 58	75 1/0	75 1/0	Plain	3.16	0.004
219	7643	--	--	470	--	49 x 30	225 1/3	150 1/3	Satin	12.90	0.011
220	--	--	--	2220	--	60 x 58	225 1/0	225 1/0	Crow	3.16	0.004
225	--	--	--	--	--	36 x 34	75 1/2	75 1/2	Plain	3.75	0.005
241	--	--	--	--	--	32 x 31	150 1/2	150 1/2	Satin	8.80	0.011
280	1675	1675	1675	--	--	42 x 30	150 1/0	150 1/0	Plain	2.82	0.004
300	--	--	--	--	--	85 x 12	900 1/0	225 1/0	Plain	3.73	0.005
325	1042	--	--	102	--	88 x 44	150 1/0	1800 1/0	Plain	0.74	0.001
328	3528	81528	81528	--	--	44 x 32	37 1/2	37 1/2	Plain	6.00	0.007
332	3732	--	--	32	--	48 x 32	37 1/0	37 1/0	Crow	12.96	0.013
333	--	--	--	--	--	48 x 32	450 1/0	450 1/0			
341	--	--	--	--	--	30 x 49	150 1/2	225 3/2	Crow	8.78	0.009
374	1674	1674	1674	--	--	40 x 30	150 1/0	150 1/0	Plain	2.82	0.004
375	1677	1677	1677	--	--	40 x 40	150 1/0	150 1/0	Plain	3.20	0.005
376	1679	--	--	--	--	44 x 36	150 1/0	150 1/0	Plain	3.20	0.004
377	1676	1676	1100	--	--	56 x 48	150 1/0	150 1/0	Plain	4.17	0.005
378	7626	7626	7626	--	--	34 x 32	75 1/0	75 1/0	Plain	5.45	0.007
401	--	--	--	807	18104	54 x 52	150 1/2	150 1/2	Crow	8.60	0.010
402	--	--	--	803	--	54 x 54	150 1/2	150 1/2	Crow	8.70	0.010
406	--	--	--	829	--	54 x 52	150 1/2	150 1/2	Twill	8.60	0.010
407	--	--	--	--	--	54 x 36	75 1/0	150 1/3	Twill	8.80	0.013
408	--	--	--	--	--	36 x 30	150 1/0	150 1/0	Plain	2.68	0.005
409	--	--	410	833	--	54 x 54	150 1/2	150 1/2	Twill	8.70	0.010
409	--	--	--	828	--	54 x 56	150 1/2	150 1/2	Twill	8.85	0.010
410	409	409	--	833	--	54 x 54	150 1/2	150 1/2	Twill	8.70	0.010
411	--	--	--	--	--	54 x 52	150 1/2	150 1/2	Twill	8.14	0.009
417	--	--	--	--	--	54 x 30	75 1/0	75 1/0	Twill 3x1		
422	--	1722	1722	817	--	48 x 22	150 2/2	31 /2	Twill	15.74	0.021
423	--	76281	76281	--	--	44 x 32	75 1/0	75 1/0	Crow	6.00	0.007
424	--	1724	1724	821	--	48 x 24	150 2/2	31 /2	Twill	16.40	0.021
432	7728	7728	--	--	--	42 x 32	75 1/0	75 1/0	Plain	6.00	0.007
434	--	--	--	--	--	40 x 26	37 1/0	37 1/0	Plain	11.28	0.013
437	--	601	601	823	--	54 x 30	150 1/2	75 1/2b	Twill	9.40	0.011
440	--	--	--	--	--	36 x 18	150 1/0	150 1/2	Leno	2.93	0.006
444	--	--	--	--	--	48 x 30	150 1/0	150 1/0	Plain	3.15	0.004
445	--	--	--	--	--	10 x 10	75 1/0	75 1/0	Plain	1.60	0.003
450	--	--	--	--	--	60 x 52	450 1/2	450 1/2	Plain	3.63	0.004
451	--	--	--	--	--	36 x 18	150 1/0	150 1/0	Leno	2.93	0.006
453	--	16453	16453	--	--	30 x 30	150 1/2	150 1/2	Plain	5.00	0.008
455	--	--	--	--	--	32 x 26	37 1/0	37 1/0	Plain	9.17	0.010
457	2180	--	--	--	--	40 x 24	225 1/0	150 1/0	Leno	2.08	0.006
466	--	--	--	--	--	15 x 15	75 2/4	75 2/4	Plain	21.76	0.031
470	7643	--	219	--	--	49 x 30	75 1/3	150 1/0	Satin	12.90	0.011
479	--	--	--	826	--	64 x 60	225 1/2	75 1/0b	Satin	8.60	0.010
483	--	604	604	814	--	42 x 30	150 2/2	150 2/2b	Twill	12.05	0.014
484	--	--	--	832	--	48 x 24	150 2/2	75 1/3b	Twill	14.60	0.018
486	--	632	632	809	--	48 x 32	150 2/2	75 1/4b	Twill	15.50	0.020
489	--	--	--	--	--	44 x 36	75 1/2	75 1/2	Twill	12.50	0.013

Weave Style	B.I.	C.S.	J.P.S.	U.M.	H.C.	Yarn Count	Construction Warp	Construction Fill	Type of Weave	Weight oz/sq-yd	Thickness in. (appx)
581	--	--	--	--	--	52 x 48	150 1/2	150 1/2	Satin	9.08	0.011
601	437	--	--	823	--	54 x 30	150 1/2	75 1/2b	Twill	9.40	0.011
604	483	--	--	814	--	42 x 30	150 2/2	150 2/2b	Twill	12.05	0.014
615	--	--	--	--	--	42 x 30	150 2/2	150 2/2b	Crow	12.05	0.014
625	--	--	--	--	--	48 x 24	150 2/2	150 2/3b	Twill	13.00	0.015
627	--	--	--	--	--	48 x 24	150 2/2	150 2/3b	Crow	13.00	0.015
630	--	--	--	--	--	48 x 22	150 2/2	150 2/4b	Twill	15.25	0.019
632	486	--	--	809	--	48 x 22	150 1/2	74 1/4b	Twill	15.50	0.020
648	--	--	--	--	--	47 x 30	37 1/0 (tex)	75 1/3			
651	--	--	--	--	--	44 x 24	37 1/0	75 1/0 (core)			
--	--	--	--	--	--			75 1/2 (tex)			
803	402	402	402	--	--	54 x 54	150 1/2	150 1/2	Crow	8.70	0.010
807	401	401	401	--	18104	54 x 52	150 1/2	150 1/2	Crow	8.60	0.010
809	486	632	632	--	--	48 x 32	150 1/2	75 1/4b	Twill	15.50	0.020
814	483	604	604	--	--	42 x 30	150 2/2	150 2/2b	Twill	12.05	0.014
817	422	1722	1722	--	--	48 x 22	150 2/2	31 /2	Twill	15.74	0.021
821	424	1724	1724	--	--	48 x 24	150 2/2	31 /2	Twill	16.40	0.021
823	437	601	601	--	--	54 x 30	150 1/2	75 1/2b	Twill	9.40	0.011
826	479	--	--	--	--	64 x 60	225 1/2	75 1/0b	Satin	8.60	0.010
828	--	--	409	--	--	54 x 56	150 1/2	150 1/2	Twill	8.85	0.010
829	406	--	--	--	--	54 x 52	150 1/2	150 1/2	Twill	8.60	0.010
830	--	--	--	--	--	48 x 24	37 1/0	75 1/3b	Twill	13.44	0.016
831	139	--	1385	--	--	64 x 60	225 1/2	225 1/2	Crow	6.64	0.007
832	484	--	--	--	--	48 x 24	150 2/2	75 1/3b	Twill	14.60	0.018
833	409	409	410	--	--	54 x 54	150 1/2	150 1/2	Twill	8.70	0.010
834	--	--	--	--	--	54 x 52	75 1/0	75 1/0	Satin	8.14	0.009
835	--	--	--	--	--	48 x 22	37 1/0	37 1/2b	Twill	14.75	0.018
836	--	--	--	--	--	64 x 24	37 1/0	75 1/3			
838	--	--	--	--	--	58 x 32	150 1/2	150 1/2	Satin	9.02	0.009
839	--	--	--	--	--	54 x 52	75 1/0	75 1/0	Twill 3x1		
840	--	--	--	--	--	54 x 30	75 1/0	75 1/2	Twill 3x1		
841	--	--	--	--	--	54 x 30	75 1/0	37 1/0	Twill 3x1		
842	--	--	--	--	--	42 x 30	37 1/0	75 1/2	Twill 3x1		
843	--	--	--	--	--	42 x 30	37 1/0	37 1/0	Twill 3x1		
844	--	--	--	--	--	48 x 24	37 1/0	75 1/3			
845	--	--	--	--	--	44 x 22	37 1/0	75 1/4			
846	--	--	--	--	--	58 x 20	150 1/2	225 1/0	Satin	5.28	0.006
850	1306	1306	1332	--	2P493	15 x 13	75 2/3	75 2/3	Plain	13.80	0.025
852	--	--	--	--	--	14 x 14	150 4/4	150 1/4	Plain	18.00	0.028
853	1314	--	1314	--	--	22 x 22	75 1/3	75 1/3	Plain	10.70	0.017
854	1311	1311	1341	--	2P499	16 x 16	75 1/3	75 1/3	Plain	7.80	0.012
855	1307	1307	1338	--	2P492	20 x 20	75 1/3	75 1/3	Plain	9.70	0.016
860	1308	1308	1330	--	2P494	12 x 12	72 2/3	72 2/3	Plain	11.30	0.020
880	1671	1680	--	80	--	72 x 70	150 1/0	150 1/0	Satin	5.70	0.006
909	--	--	--	--	--				Hi Mod	9.50	0.010
918	--	--	--	--	--				Hi Mod	18.20	0.018
990	--	7590	--	--	--	10 x 5	75 2/3	75 2/3	Leno	9.26	0.020
991	1505	1505	1505	--	1505	10 x 10	75 2/3	75 2/3	Plain	9.90	0.018
992	--	7589	--	--	--	13 x 12	75 2/3	75 2/3	Plain	12.50	0.018
994	--	--	--	--	--	10 x 5	150 3/2	150 3/4	Leno	4.98	0.009
1000	1500	1500	1500	--	--	16 x 14	150 4/2	150 4/2	Plain	9.66	0.014
1022	1533	1533	1533	--	--	18 x 18	150 2/2	150 2/2	Plain	5.79	0.009

Weave Style	B.I.	C.S.	J.P.S.	U.M.	H.C.	Yarn Count	Construction Warp	Construction Fill	Type of Weave	Weight oz/sq-yd	Thickness in. (appx)
1042	--	325	325	102	--	88 x 44	900 1/0	1800 1/0	Plain	0.74	0.001
1044	1544	1544	1544	--	--	28 x 14	150 4/2	150 4/2	Basket	18.00	0.022
1061	--	--	--	--	--	62 x 52	900 1/0	900 1/0	Plain	0.75	0.001
1070	--	--	--	--	--	60 x 35	450 1/0	900 1/0	Plain	1.04	0.002
1071	--	--	--	--	--	60 x 30	900 1/0	900 1/0	Plain	0.64	0.001
1073	--	--	--	--	--	60 x 30	450 1/0	900 1/0	Plain	0.96	0.001
1080	--	--	--	--	--	60 x 47	450 1/0	450 1/0	Plain	1.43	0.002
1100	1676	1676	--	377	--	56 x 48	150 1/0	150 1/0	Plain	4.17	0.005
1116	--	--	--	--	--	60 x 58	450 1/2	450 1/2	Plain	3.19	0.004
1125	--	--	--	--	--	40 x 39	450 1/2	450 1/2	Plain	2.60	0.004
1131	1211	--	--	--	--	120 x 52	450 1/0	150 1/0	Plain	3.36	0.005
1134	--	--	--	--	--	120 x 44	450 1/0	450 1/0	Plain	3.23	0.004
1161	--	--	--	--	--	100 x 42	450 1/0	100 1/0			
1165	--	--	--	--	--	60 x 52	450 1/2	150 1/0	Plain	3.66	0.004
1205	--	--	--	--	--	60 x 52	450 1/2	150 1/0	Satin	3.65	0.004
1211	--	1131	1131	--	--	120 x 52	450 1/0	150 1/0	Plain	3.36	0.005
1212	--	--	--	--	--	40 x 39	450 1/2	150 1/0	Plain	2.66	0.004
1278	--	--	--	--	--	20 x 20	450 1/2	450 1/2	Plain	1.08	0.003
1289	--	--	--	--	--	50 x 40	450 1/2	450 1/2	Plain	1.16	0.002
1298	--	--	--	--	--	50 x 20	450 1/0	225 1/0	Plain	1.22	0.003
1299	--	--	--	209	--	50 x 20	450 1/0	450 1/0	Plain	0.94	0.002
1299	--	--	--	--	--	50 x 20	450 1/0	900 1/0			
1302	--	1310	--	--	2P497	15 x 15	75 2/4	75 2/4	Plain	20.50	0.031
1305	--	--	1334	--	2P490	15 x 14	75 2/3	75 2/3	Plain	15.50	0.031
1306	--	--	1332	850	2P493	14 x 13	75 2/3	75 2/3	Plain	14.00	0.030
1307	--	--	1338	855	2P492	20 x 20	75 1/3	75 1/3	Plain	9.70	0.015
1308	--	--	1330	860	2P494	12 x 12	75 2/3	75 2/3	Plain	11.30	0.028
1310	--	--	1302	--	2P497	15 x 15	75 2/4	75 2/4	Plain	20.50	0.031
1311	--	--	1341	854	2P499	16 x 16	75 1/3	75 1/3	Plain	8.00	0.012
1311	--	1391	--	--	--	13 x 12	75 2/4	75 2/4	Plain	16.50	0.028
1312	--	--	1336	--	2P496	20 x 18	75 1/3	75 1/3	Plain	9.50	0.014
1312	--	1317	--	--	2P507	17 x 17	75 1/3	75 1/3	Plain	8.30	0.014
1314	--	--	--	853	--	22 x 22	75 1/3	75 1/3	Plain	10.70	0.017
1317	--	--	1312	--	2P507	17 x 17	75 1/3	75 1/3	Plain	8.30	0.014
1319	--	--	--	--	--	14 x 13	75 1/4	75 1/4	Plain	12.10	0.018
1322	--	--	1340	--	2P503	22 x 22	75 1/3	75 1/3	Plain	11.00	0.018
1330	1308	1308	--	860	2P494	12 x 12	75 2/3	75 2/3	Plain	11.30	0.028
1332	1306	1306	--	850	2P493	14 x 13	75 2/3	75 2/3	Plain	14.00	0.030
1334	1305	1305	--	--	2P490	15 x 14	75 2/3	75 2/3	Plain	15.50	0.031
1335	--	--	--	--	--	15 x 15	75 2/3	75 2/3	Plain	14.08	0.028
1336	1312	1312	--	--	2P496	20 x 18	75 1/3	75 1/3	Plain	9.50	0.014
1338	1307	1307	--	855	2P492	20 x 20	75 1/3	75 1/3	Plain	9.70	0.015
1339	--	--	--	--	--	20 x 21	75 1/3	75 1/3	Plain	9.63	0.014
1340	--	1322	--	--	2P503	22 x 22	75 1/3	75 1/3	Plain	11.00	0.018
1341	1311	1311	--	854	2P499	16 x 16	75 1/3	75 1/3	Plain	8.00	0.012
1385	139	--	--	831	--	64 x 60	225 1/2	225 1/2	Crow	6.64	0.007
1391	--	--	1311	--	--	13 x 12	75 2/4	75 2/4	Plain	16.50	0.028
1418	--	--	--	--	--	60 x 52	450 1/2	150 1/0b	Plain	3.79	0.006
1419	--	--	--	--	--	62 x 60	225 1/2	150 1/2b	Satin	8.70	0.010
1428	--	--	--	--	--	44 x 32	150 1/2	150 1/2b	Plain	6.25	0.008
1437	--	--	--	--	--	32 x 32	dacron	75 1/0	Plain	3.14	0.005
1456	--	--	--	5614	--	32 x 28	dacron	150 1/0	Plain	1.90	0.004

Weave Style	B.I.	C.S.	J.P.S.	U.M.	H.C.	Yarn Count	Construction		Type of Weave	Weight oz/sq-yd	Thickness in. (appx)
							Warp	Fill			
1504	--	--	--	--	--	20 x 10	75 1/3	75 1/3	Leno	9.90	0.019
1505	--	--	--	991	--	10 x 10	75 2/3	75 2/3	Plain	9.90	0.018
1510	--	--	--	--	--	32 x 39	150 1/2	150 1/2	Plain	4.88	0.006
1520	--	--	--	1542	--	18 x 18	150 3/2	150 3/2	Plain	8.46	0.012
1522	--	--	--	122	--	24 x 22	150 1/2	150 1/2	Plain	3.69	0.005
1523	--	--	--	--	--	28 x 20	150 3/2	450 1/2	Plain	11.85	0.014
1523	--	--	--	--	--	28 x 16	150 3/0	150 3/0	Plain	7.30	0.010
1524	--	--	--	--	--	26 x 26	150 1/2	150 1/2	Plain	5.30	0.007
1525	--	--	--	--	--	30 x 24	150 1/2	150 1/2	Plain	4.50	0.006
1526	--	--	--	--	--	34 x 32	150 1/3	150 1/3	Plain	5.45	0.007
1527	--	--	--	--	--	17 x 17	150 3/2	150 3/2	Plain	12.90	0.017
1528	--	--	--	--	--	42 x 32	150 1/2	150 1/2	Plain	6.00	0.007
1529	--	--	--	--	--	42 x 28	150 1/3	150 1/0	Plain	4.30	0.005
1530	1566	--	--	--	--	19 x 18	150 3/3	150 3/3	Plain	13.30	0.017
1532	--	--	--	1032	--	16 x 14	150 3/2	150 3/2	Plain	7.23	0.010
1533	1561	1561	1561	--	1561	16 x 14	150 3/3	150 3/3	Plain	10.85	0.016
1533	--	--	--	1033	--	18 x 18	75 2/2	75 2/2	Plain	5.79	0.009
1534	--	--	--	--	--	16 x 14	150 2/2	75 2/3	Plain	12.00	0.016
1539	--	--	--	--	--	39 x 28	150 1/2	150 1/2	Plain	5.50	0.007
1542	1520	1520	1520	--	--	18 x 18	150 3/2	150 3/2	Plain	8.46	0.012
1543	--	--	--	--	--	49 x 30	150 2/2	450 1/2	Crow	8.78	0.009
1544	--	--	--	1044	--	28 x 14	150 4/2	150 4/2	Basket	18.00	0.022
1548	1587	1587	1587	--	--	42 x 21	150 4/2	150 4/2	Leno	20.80	0.028
1557	--	--	--	--	--	57 x 30	150 1/2	450 1/2	Crow	5.42	0.006
1559	--	--	--	--	--	59 x 30	150 1/2	225 1/0	Crow	5.55	0.006
1561	--	--	--	1533	--	16 x 14	150 3/3	150 3/3	Plain	10.85	0.016
1562	--	--	--	194	--	30 x 16	150 1/0	150 1/0	Leno	1.93	0.005
1564	--	--	--	--	--	20 x 18	150 4/2	150 4/2	Plain	12.70	0.016
1566	--	--	--	1530	--	19 x 18	150 3/3	150 3/3	Plain	13.30	0.017
1567	--	--	--	--	--	60 x 20	150 1/2	225 1/0	Crow	5.18	0.005
1573	1577	--	--	1577	--	49 x 20	150 2/2	450 1/2	Crow	8.64	0.007
1573	--	1573	--	--	--	60 x 20	150 1/2	450 1/2	Crow	5.18	0.005
1577	--	--	--	--	--	60 x 20	150 1/2	450 1/2	Crow	5.18	0.005
1581	--	--	--	--	--	57 x 54	150 1/2	150 1/2	Satin	8.92	0.009
1582	--	--	--	--	--	60 x 56	150 1/3	150 1/3	Satin	13.45	0.014
1583	--	--	--	--	--	54 x 48	150 2/2	150 2/2	Satin	16.75	0.018
1584	--	--	--	--	--	42 x 36	150 4/2	150 4/2	Satin	25.40	0.027
1587	--	--	--	1548	--	42 x 21	150 4/2	150 4/2	Leno	20.80	0.028
1588	--	--	--	--	--	42 x 35	150 4/4	150 4/4	Satin	51.80	0.050
1589	--	--	--	--	--	13 x 12	150 4/3	150 4/3	Plain	12.50	0.018
1590	--	--	--	--	--	10 x 5	150 4/3	150 4/5	Leno	9.26	0.020
1597	--	--	--	3000	--	30 x 30	75 2/4	75 2/4	Fancy	38.50	0.045
1598	--	--	--	--	--	16 x 8	150 3/2	150 3/2	Leno	8.05	0.019
1609	--	--	--	--	--	32 x 10	150 1/0	150 1/0			
1610	--	--	--	192	--	32 x 28	150 1/0	150 1/0	Plain	2.48	0.004
1611	--	--	--	3216	--	32 x 16	150 1/0	75 1/0	Plain	2.48	0.004
1614	--	--	--	--	--	30 x 14	150 1/0	75 1/0	Plain		
1619	--	--	--	--	--	20 x 10	150 1/0	75 1/0	Plain	1.60	0.004
1620	--	--	--	196	--	20 x 20	150 1/0	150 1/0	Plain	1.60	0.004
1621	--	196	196	--	--	30 x 14	150 1/0	150 1/0	Leno	2.36	0.006
1624	--	--	--	--	--	26 x 24	150 1/0	150 1/0	Plain	1.84	0.004
1626	--	--	--	--	--	40 x 26	150	75 1/0			

Weave Style	B.I.	C.S.	J.P.S.	U.M.	H.C.	Yarn Count	Construction		Type of Weave	Weight oz/sq-yd	Thickness in. (appx)
							Warp	Fill			
1653	--	--	--	--	--	16 x 8	150 1/0	75 1/0	Plain	1.40	0.005
1654	--	--	--	--	--	20 x 8	150 1/0	40 /1	Plain	2.00	0.008
1655	--	--	--	--	--	10 x 5	150 1/0	150 1/0	Leno	0.46	0.003
1656	--	--	--	--	--	30 x 16	150 1/0	75 1/0	Leno	2.57	0.005
1658	--	--	--	198	--	20 x 10	150 1/0	75 1/0	Plain	1.60	0.004
1659	--	--	--	203	--	20 x 10	150 1/0	75 1/0	Leno	1.60	0.004
1660	--	--	--	--	--	60 x 12	150 1/0	150 1/0	Plain	2.80	0.005
1663	--	--	--	--	--	50 x 8	150 1/0	450 1/0			
1664	--	--	--	--	--	64 x 8	150 1/0	900 1/0	Plain	2.60	0.002
1666	--	--	--	--	--	60 x 10	150 1/0	225 1/0			
1667	--	--	--	--	--	60 x 12	150 1/0	900 1/0	Plain	2.50	0.004
1668	--	--	--	208	--	60 x 10	150 1/0	225 1/0	Plain	2.86	0.004
1668	--	--	--	--	--	20 x 6	150 1/0	75 1/0	Leno	1.23	0.002
1671	--	1680	880	80	--	72 x 70	150 1/0	150 1/0	Satin	5.70	0.006
1674	--	--	--	--	--	40 x 32	150 1/0	150 1/0	Plain	2.83	0.004
1675	--	--	--	280	--	40 x 32	150 1/0	150 1/0	Plain	2.82	0.004
1676	--	--	1100	377	--	56 x 48	150 1/0	150 1/0	Plain	4.17	0.005
1677	--	--	--	375	--	40 x 40	150 1/0	150 1/0	Plain	3.20	0.005
1679	--	--	--	376	--	44 x 36	150 1/0	150 1/0	Plain	3.20	0.005
1680	1671	--	880	80	--	72 x 70	150 1/0	150 1/0	Satin	5.70	0.006
1681	--	--	--	--	--	56 x 36	150 1/0	150 1/0	Plain	3.60	0.005
1683	--	--	--	--	--	8 x 3	150 1/0	75 1/0	Leno	0.54	0.001
1686	--	--	--	--	--	20 x 6	150 1/0	75 1/0	Leno	1.23	0.002
1695	--	--	--	--	--	40 x 24	150 1/0	75 1/0			
1722	422	--	--	817	--	48 x 22	150 2/2	31 /2	Twill	15.74	0.021
1724	424	--	--	821	--	48 x 24	150 2/2	31 /2	Twill	16.40	0.024
1800	--	--	--	--	--	16 x 14	18 1/0	18 1/0	Plain	9.60	0.014
1884	--	--	--	--	--	44 x 36	18 1/0	18 1/0	Satin	25.60	0.028
1885	--	--	--	--	--	64 x 20	18 1/0	25 1/0	Satin	25.09	0.026
1912	117	121	121	--	--	54 x 39	450 1/2	450 1/2	Plain	2.49	0.003
1925	--	--	--	--	--	22 x 15	37 1/3	37 1/3	Plain	18.50	0.030
1951	--	84205	84205	4205	--	20 x 14	150 2/3b	150 2/3b	Plain	9.30	0.017
1958	--	--	--	--	--	20 x 16	18 1/0	18 1/0	Plain	13.00	0.020
1978	--	--	--	--	--	18 x 14	25 1/0	25 1/0	Plain	8.70	0.010
2112	--	--	--	--	--	40 x 39	225 1/0	225 1/0	Plain	2.11	0.003
2113	--	--	--	--	--	60 x 56	225 1/0	450 1/0	Plain	2.38	0.004
2116	2185	216	--	--	--	60 x 58	225 1/0	225 1/0	Plain	3.16	0.004
2120	--	--	--	--	--	20 x 20	225 1/0	225 1/0		1.08	0.003
2121	--	--	--	--	--	20 x 20	225 1/0	450 1/0	Plain		
2125	--	--	--	--	--	40 x 39	225 1/0	150 1/0	Plain	2.60	0.004
2165	--	--	--	--	--	60 x 52	225 1/0	150 1/0	Plain	3.66	0.004
2180	--	--	--	457	--	40 x 24	225 1/0	150 1/0	Leno	2.08	0.006
2185	--	216	2116	--	--	60 x 58	225 1/0	225 1/0	Plain	3.16	0.004
2220	--	220	220	--	--	60 x 58	225 1/0	225 1/0	Crow	3.16	0.004
2520	--	--	--	2542	--	18 x 18	25 1/0	25 1/0	Plain	8.46	0.012
2523	--	--	--	--	--	28 x 20	25 1/0	25 1/0	Plain	11.85	0.014
2532	--	--	--	--	--	16 x 14	25 1/0	25 1/0	Plain	7.23	0.010
2542	2520	2520	2520	--	--	18 x 18	25 1/0	25 1/0	Plain	8.46	0.012
3000	1597	1597	1597	--	1597	30 x 30	75 2/4	75 2/4	Fancy	38.50	0.045
3005	7597	--	--	--	--	27 x 27	75 2/4	75 2/4	Fancy	34.80	0.045
3128	--	31528	31528	3528	--	42 x 32	150 1/2	150 1/2	Crow	6.00	0.007
3138	--	--	--	--	--	64 x 60	450 2/2	450 2/2	Crow	6.64	0.007

Weave Style	B.I.	C.S.	J.P.S.	U.M.	H.C.	Yarn Count	Construction Warp	Construction Fill	Type of Weave	Weight oz/sq-yd	Thickness in. (appx)
3228	--	--	--	--	--	32 x 28	37 1/0	37 1/0	Plain	9.30	0.011
3264	--	--	--	--	--	46 x 42	150 1/2	150 1/2	Plain	7.42	0.008
3281	--	--	--	--	--	15 x 13.5	75 1/3	75 1/3	Knit	8.00	0.015
3292	1632	1632	1632	--	--	32 x 32	150 1/0	75 1/0	Plain	3.78	0.005
3415	--	--	--	--	--	30 x 49	450 1/2	150 1/2	Crow	8.78	0.009
3468	--	--	--	--	--	70 x 58	150 1/0	150 1/0	Twill	5.13	0.006
3528	3128	31528	31528	--	--	42 x 32	150 1/2	150 1/2	Crow	6.00	0.007
3528	--	81528	81528	328	--	42 x 32	150 1/2	150 1/2	Plain	6.00	0.007
3529	--	--	--	--	--	60 x 52	150 1/3	150 1/3	Satin	12.45	0.013
3601	--	--	--	--	--	53 x 27	75 1/0	37 1/0	Twill 3x1		
3602	--	--	--	--	--	53 x 30	75 1/0	150 1/0 (core)	Twill 3x1		
--	--	--	--	--	--			50 1/0 (tex)			
3604	--	--	--	--	--	42 x 30	37 1/0	75 1/2	Twill 3x1		
3625	--	--	--	--	--	48 x 24	37 1/0	75 1/3	Twill 3x1		
3632	--	--	--	--	--	48 x 22	37 1/0	75 1/4	Twill 2x2b		
3700	--	--	--	--	--	16 x 14	37 1/2	37 1/2	Plain	9.66	0.014
3701	--	--	--	--	--	12 x 6	37 1/0	18 1/0			
3722	3733	3733	3733	--	3733	18 x 18	37 1/0	37 1/0	Plain	5.79	0.009
3732	--	332	332	32	--	48 x 32	37 1/0	37 1/0	Crow	12.96	0.013
3733	--	--	--	3722	--	18 x 18	37 1/0	37 1/0	Plain	5.79	0.009
3743	--	--	--	--	--	49 x 30	37 1/0	225 1/0	Crow	8.78	0.009
3783	--	--	--	--	--	54 x 48	37 1/0	37 1/0			
3784	--	--	--	--	--	42 x 36	37 1/0	37 1/0			
4032	--	--	--	--	--	55 x 8	150 1/0	450 1/0	Plain	2.21	0.004
4068	--	--	--	--	--	40 x 29	150 1/0	150 1/0	Plain	2.65	0.005
4071	--	--	--	--	--	26 x 26	75 2/4b	75 2/4b	Fancy	34.00	0.044
4081	--	--	--	--	--	32 x 8	150 1/0	450 1/0	Plain	1.43	0.003
4205	1951	84205	84205	--	--	20 x 14	150 2/3b	150 2/3b	Plain	9.30	0.017
4209	--	--	--	--	--	20 x 14	75 1/3b	75 1/3b	Basket	8.33	0.010
4212	--	--	--	--	--	22 x 16	75 1/4b	75 1/4b	Basket	12.44	0.016
4700	--	--	--	--	--	14 x 13	37 1/0	37 1/0			
5614	1456	--	--	--	--	32 x 28	Dacron	150 1/0	Plain	1.90	0.004
6015	--	--	--	--	--	53 x 30	150 1/2	75 1/2	Twill 3x1		
6045	--	--	--	--	--	42 x 30	75 1/2	75 1/2	Twill 3x1		
6257	--	--	--	--	--	49 x 24	37 1/0	37 1/0b	Twill	11.60	0.014
6275	--	--	--	--	--	48 x 24	75 1/2	75 1/3	Twill 3x1		
6324	--	--	--	--	--	44 x 22	75 1/2	75 1/4	Twill 2x2b		
6328	--	--	--	--	--	48 x 22	75 1/2	75 1/4	Twill 2x2b		
7500	--	--	--	--	--	16 x 14	75 2/2	75 2/2	Plain	9.66	0.014
7520	--	--	--	7542	--	18 x 18	75 1/3	75 1/3	Plain	8.46	0.012
7522	7533	7533	7533	--	7533	18 x 18	75 1/2	75 1/2	Plain	5.79	0.009
7523	--	--	--	--	--	28 x 20	75 1/3	75 1/3	Plain	11.70	0.014
7532	--	--	--	--	--	16 x 14	75 1/3	75 1/3	Plain	7.23	0.010
7533	--	--	--	7522	--	18 x 18	75 1/2	75 1/2	Plain	5.79	0.009
7534	7634	--	--	7634	--	16 x 14	75 2/2	75 2/3	Plain	12.00	0.016
7539	7630	7630	--	--	--	38 x 28	75 1/0	75 1/0	Plain	5.50	0.006
7542	7520	7520	7520	--	--	18 x 18	75 1/3	75 1/3	Plain	8.46	0.012
7543	--	--	--	--	--	49 x 30	75 1/2	450 1/2	Crow	8.78	0.009
7544	--	--	--	--	--	28 x 14	75 2/2	75 4/2	Basket	18.00	0.022
7548	7687	7587	7587	--	7587	42 x 21	75 2/2	75 2/2	Leno	20.80	0.028
7549	--	--	--	--	--	30 x 28	75 3/0b	75 3/0b	Fancy	14.22	0.015
7557	7657	7657	--	--	--	57 x 30	75 1/0	225 1/0	Crow	5.42	0.006

Weave Style	B.I.	C.S.	J.P.S.	U.M.	H.C.	Yarn Count	Construction Warp	Fill	Type of Weave	Weight oz/sq-yd	Thickness in. (appx)
7576	--	--	--	--	--	120 x 24	75 1/0	150 1/0	Satin	10.34	0.012
7581	7781	7781	7781	--	--	57 x 54	75 1/0	75 1/0	Satin	8.92	0.009
7585	--	--	--	--	--	64 x 20	75 2/2	75 1/3	Satin	25.40	0.027
7586	--	--	--	--	--	62 x 20	75 2/2	75 1/3	Satin	25.00	0.027
7587	7687	--	--	7548	--	42 x 21	75 2/2	75 2/2	Leno	20.80	0.028
7589	--	--	--	992	--	13 x 12	75 3/2	75 3/2	Plain	12.50	0.018
7590	--	--	--	990	--	10 x 5	75 2/3	75 2/3	Leno	9.26	0.020
7597	--	--	--	3000	--	30 x 30	75 2/4	75 2/4	Fancy	38.50	0.045
7603	--	--	--	--	--	10 x 5	75 1/3	75 2/3	Leno	4.71	0.012
7605	--	--	--	--	--	7.5 x 7.5	75 2/5	75 2/5	Plain	13.38	0.020
7606	--	--	--	--	--	6/6 x 6	75 2/3	75 2/3	Plain	5.96	0.013
7607	--	--	--	--	--	5/5 x 5	75 2/5	75 2/5	Plain	8.06	0.019
7608	--	--	--	--	--	7/7 x 7	75 2/5	75 2/5	Plain	11.54	0.020
7609	--	--	--	--	--	8/8 x 8	150 3/3	150 3/3	Plain	6.20	0.015
7610	--	--	--	--	--	10/5 x 10	75 2/5	75 2/5	Basket	16.08	0.027
7611	--	--	--	--	--	8/8 x 8	75 2/4	75 2/4	Plain	10.93	0.019
7612	--	--	--	--	--	16 x 8	75 1/3	75 2/3	Leno	8.08	0.015
7613	--	--	--	--	--	10 x 5	75 2/3	75 2/5	Leno	9.54	0.025
7614	--	--	--	--	--	6/6 x 6	75 2/5	75 2/5	Plain	10.14	0.020
7615	--	--	--	--	--	14/14x 14	75 1/2	75 1/2	Plain	5.09	0.009
7617	--	--	--	--	--	48 x 30	75 1/0	75 1/0	Twill	6.18	0.006
7621	--	--	--	--	--	26 x 22	75 1/0	75 1/0	Plain	3.93	0.005
7626	--	--	--	378	--	34 x 32	75 1/0	75 1/0	Plain	5.45	0.007
7628	--	--	--	7629	--	44 x 32	75 1/0	75 1/0	Plain	6.00	0.007
7629	7628	7628	7628	--	--	44 x 32	75 1/0	75 1/0	Plain	6.00	0.007
7630	--	--	7539	7639	--	38 x 28	75 1/0	75 1/0	Plain	5.50	0.006
7634	--	7534	7534	--	7534	16 x 14	75 2/2	75 2/3	Plain	12.00	0.016
7637	--	--	--	--	--	44 x 22	75 1/0	37 1/0	Plain	6.80	0.008
7639	7630	7630	--	--	--	39 x 28	75 1/0	75 1/0	Plain	5.50	0.007
7641	--	--	--	--	--	32 x 21	75 1/2	75 1/2	Plain	8.50	0.009
7642	--	--	--	--	--	42 x 20	75 1/0	37 (lex)	Plain	6.85	0.010
7643	--	--	219	470	--	49 x 30	75 1/3	150 1/0	Satin	12.90	0.011
7645	--	--	--	--	--	45 x 42	75 1/2	75 1/2	Plain	13.36	0.016
7651	--	--	--	--	--	10/10x 10	75 2/3	75 2/3	PLain	10.30	0.019
7657	--	--	7557	7557	--	57 x 30	75 1/0	225 1/0	Crow	5.42	0.006
7661	--	--	--	--	--	84 x 20	75 2/2	150 1/0	Satin	27.80	0.032
7664	--	--	--	--	--	20 x 18	75 2/2	75 2/2	Plain	12.70	0.016
7677	--	--	--	--	--	34 x 32	75 1/0	75 1/0	Plain	5.07	0.006
7684	--	--	--	--	--	42 x 32	75 2/2	75 2/2	Satin	25.40	0.027
7687	--	7587	7548	7587	--	42 x 21	75 2/2	75 2/2	Leno	20.80	0.028
7692	--	--	--	--	--		75 1/0	37			
7728	--	--	--	432	--	42 x 32	75 1/0	75 1/0	Plain	6.00	0.007
7743	--	--	--	7575	--	120 x 20	75 1/0	150 1/0	Satin	10.20	0.011
7781	--	--	--	7581	--	57 x 54	75 1/0	75 1/0	Satin	8.92	0.009
7784	--	--	--	--	--	44 x 35	75 2/2	75 2/2	Satin	25.50	0.026
7787	--	--	--	--	--	42 x 21	18 1/0	18 1/0	Leno	20.16	0.028
8000	--	--	--	--	--	80 x 8	150 1/2	Polyester	Plain		
8001	--	--	--	--	--	60 x 24	450 1/2	75 1/2	Plain	5.50	0.012
8010	--	--	--	--	--					26.20	0.029
8116	--	--	--	--	--	60 x 32	150 1/0	150 1/0	Plain	3.02	0.004
8430	--	15843	15843	--	--	44 x 30	150 4/2	150 1/0	Crow	15.90	0.019
8584	--	--	--	--	--	44 x 35	150 4/2	150 4/2	Satin	25.50	0.026

Weave Style	B.I.	C.S.	J.P.S.	U.M.	H.C.	Yarn Count	Construction Warp	Construction Fill	Type of Weave	Weight oz/sq-yd	Thickness in. (appx)
9180	--	--	--	--	--	8/4 x 8	18 1/0	18 1/0	Plain	5.51	0.020
15811	--	--	--	--	--	57 x 54	150 1/0	150 1/0	Satin	8.92	0.009
15843	--	--	--	8430	--	44 x 30	150 4/2	150 1/0	Crow	15.90	0.019
16453	--	--	--	543	--	30 x 30	150 1/2	150 1/2	Plain	5.00	0.008
31528	3128	--	--	3528	--	42 x 32	150 1/2	150 1/2	Crow	6.00	0.007
76281	--	--	--	423	--	42 x 32	75 1/0	75 1/0	Crow	6.00	0.007
81528	3528	--	--	328	--	42 x 32	150 1/2	150 1/2	Plain	6.00	0.007
81584	--	--	--	--	--	42 x 36	150 4/2	150 4/2	Satin	25.40	0.027
84205	1951	--	--	4205	--	20 x 14	150 2/3b	150 2/3b	Plain	9.30	0.017
84207	--	--	--	--	--						
87628	--	--	--	--	--	44 x 32	75 1/0	75 1/0	Plain	6.00	0.007

APPENDIX 3

Table 3-1. List of Kevlar Weave Styles

Dup = Dupont
CS = Clark Schwebel
Hex = Hexcel
BI = Burlington Industri
JPS = J.P. Stevens
TPI = Textile Products, Inc.

KEVLAR 49 FABRICS

Dup	CS	Hex	BI	TPI	JPS	Yarn Count	Const/Denier Warp	Fill	Const/Tensile Warp	Fill	Type of Weave	Weight oz/sq.yd	Thickness (mils)
				7247		60 X 60	55	55			Plain	0.9	0.002
				7645		34 X 34	195	195			Plain	1.44	0.003
120	350	120	5120		7314	34 X 34	195	195	265	250	Plain	1.7	4.5
220	351		5220			22 X 22	380	380	300	300	Plain	2.2	4.5
				7646		17 X 17	1140	1140			CFS	4.49	0.009
500	354	500	5500			13 X 13	1420	1420	610	620	Plain	4.8	9.0
				7378		17 X 17	1140	1140			Plain	4.93	0.010
				7372		50 X 50	380	380			8HSW	4.93	0.009
				7613		13 X 13	1420	1420			Plain	4.93	0.010
181	348	181	5181		7318	50 x 50	380	380	630	625	Satin	5.0	9.0
281	352	281	5281		7312	17 X 17	1140	1140	620	655	Plain	5.1	10.0
285	353	285	7315		7315	17 X 17	1140	1140	630	645	Crow	5.1	10.1
143	343	143	5143			100 X 20	380	195	1470	155	Crow	5.6	12.0
243	358		5243			38 X 18	1140	380	1430	260	Crow	6.6	12.0
328	328		5328			17 X 17	1420	1420	745	755	Plain	6.3	13.0
335	323		5335			17 X 17	1420	1420	745	755	Crow	6.2	12.0
900						17 X 17	2130	2130			Satin	9.0	13.0
1050	384	1050	5150			28 X 28	1420	1420	1350	1350	Basket 4X4	10.5	18.0
1350	386	1350	5350			26 X 22	2130	2130			Basket 4 X 4	13.5	29.0
	1492					33 X 33	1420	1420			Basket 2 X 2	14.3	25.0
1033	388		5133			40 X 40	1420	1420	1900	1910	Basket 8 X 8	16.0	31.0

KEVLAR 29 FABRICS

Dup	CS	Hex	BI	JPS	Yarn Count	Const/Denier Warp	Fill	Const/Tensile Warp	Fill	Type of Weave	Weight oz/sq.yd	Thickness (mils)
	740			7306	44 X 44	200	200	295	295	Plain	2.0	5.0
	732			7308	40 X 40	200	200	290	290	Plain	2.0	5.0
				7310	22 X 22	1000	1000	700	700	Plain	5.8	10.0
713	713	713	26434	7305	31 X31	1000	1000	850	850	Plain	8.0	15.0
710	710	710	26032	710	24 X 24	1500	1500	1100	1200	Plain	9.4	17.0
720	728	728		718	17X17	1500	1500	800	850	Plain	6.8	12.0
	735		26036	7316	34 X 34	1500	1500	1500+	1500+	Basket 2 X 2	14.5	25.0
			26435		39 X 34	1500	1500	2100+	2100+	Basket 2 X 2	16.2	27.0
				7300	42 X 42	1500	1500	2000+	2000+	Basket 7 X 7	17.0	31.0
	748				48 X 48	1500	1500	2500+	2500+	Basket 8 X 8	19.7	40.0

APPENDIX 4

NO MORE RIVETS

Note: *This Article originally appeared in "Homebuilt Aircraft" magazine in the January 1984 issue, and is reproduced by permission of the author, Mr. Otis Holt.*

Adhesive bonding can be an attractive means of connecting aluminum aircraft components. Among its attributes are economy, ease of assembly, the transfer of loads through large areas, smooth resulting surfaces and great strength. For these reasons, structural bonding of aluminum has become common in the aerospace industry, where bonded assemblies are routinely subjected to the most severe extremes of stress and environment. However, the path to this happy state of affairs has been strewn with problems, and there are pitfalls that the homebuilder should be aware of before hastening to glue his or her metal airplane together.

My experience with the subject began during Oshkosh '82, a mere two days after I'd parted with the price of my homebuilt Moni kit, which requires the bonding of aluminum wing skins to aluminum ribs and spars. I learned that Andrew C. Marshall, a leading expert and industry consultant on all types of bonded sandwich structures from Walnut Creek, California, is critical of amateur aluminum bonding in general and of the technique specified for my kit in particular. Marshall had been the featured speaker some months earlier at our EAA Chapter 124 meeting, and his speech had left me with little doubt of his knowledge. Marshall was available for consultation at Oshkosh '82 through the sponsorship of **Western Flyer** and **Ultralight Flyer** magazines at their ARV/Ultralight design contest tent.

The technique prescribed for bonding my wings consists of thorough scuffing the bond area using an abrasive disk, meticulously wiping the surface with Methyl Ethyl Ketone, and bonding directly to the bare aluminum using a room temperature curing two-part epoxy, (Hysol EA9410). Let's call this the scuff/wipe method. Marshall assured me that, while such bonds may initially appear quite strong, they are not reliable. In fact, he claimed that, given **sufficient time and environmental exposure**, (particularly any form of moisture), delamination is inevitable. Although I'm always wary of such absolute statements, Marshall's credentials and a powerful sense of self-preservation prevented me from taking this one lightly.

I thus resolved to study the problem to either vindicate the scuff/wipe method or possibly find a more suitable technique. I've since accumulated a pile of formidable published studies, along with several product date-sheets and, most importantly, the written and verbal opinions of many prominent people in the aerospace industry. Little of this has contradicted Marshall's opinion, with one notable exception on which a discussion follows.

It soon became evident that a gap exists between most of the homebuilt fraternity and a vast body of knowledge and experience on the subject in the aerospace industry. The bulk of this knowledge is the direct result of a Herculean effort to overcome problems inherent in aluminum bonding. This article will provide a modest start at bridging the gap and help prevent the repetition of what are now some old mistakes.

THE PROBLEM

To understand the problem, you must realize that what you actually bond to is not the aluminum itself, but rather the ever-present oxide layer on the surface. Even if you grind away the existing layer, rapid oxidation begins immediately upon exposure of the new surface. The routine objective of all modern aluminum bonding surface preparation is to develop a consistent, uniform oxide film that is:

1) Adhesively strong in attachment to the base metal,
2) Cohesively strong within itself,
3) Wettable by the adhesive,
4) Active in terms of mechanical keying and/or the formation of chemical bonds, and
5) Environmentally stable (*1).

Industry experience has shown that the neglect of any of these criteria will seriously affect bond strength and/or durability. While it might naturally be hoped that the lesser requirements of homebuilts would permit some compromise of the above, this does not appear to be the case. As we will see, disbond problems experienced by the industry cannot be related to any special service conditions that would preclude homebuilding.

The scuff/wipe method is somewhat deficient in meeting goals number 1, 2 and 3, and **seriously** deficient with regard to number 5. In a study conducted in 1975 by Murray Kuperman of United Airlines, test specimens prepared using a similar method and exposed stress-free to 95% relative humidity at 95° F for just 30 days retained less than 50% of the shear strength of identical control specimens that were kept dry during the same period. The results of this study vividly illustrate the relationship between surface preparation and bond durability.

Oxides that form during uncontrolled exposure to the air and humidity are poorly attached to the base metal and also lack cohesive strength, so even the initial strength of a bond is limited by oxides' presence.

But Marshall, and others, have verified that even if the oxide layer in a bare metal bond were non-existent or of acceptable quality initially, condensation and humidity exposure would result in oxide formation and/or degradation beginning at the unprotected edges of the joint and migrating across the bond interface, eventually resulting in complete delamination. Whether failure would occur within the service life of a particular aircraft is impossible to say because it depends upon initial cleanliness, environment, the particular adhesive used and many other factors. (*6).

Although vibration and other stresses can accelerate deterioration, it is primarily dependent upon the presence of moisture, even in minute amounts, high humidity is the worst culprit (*3). This gives rise to the possibility that aircraft containing bare metal bonds could go from sound to unsafe during even a storage period, due to nothing more than temperature/humidity fluctuations, with little visible evidence to warn the pilot. Slosh-priming after assembly might help to isolate the bond, but difficulty in assuring adequate adhesion and thorough coverage in complex assemblies, without adding excess weight, makes this technique less than satisfactory. Also, expansion/contraction differences between the components of an assembly could cause cracks in the prime coat—where you least want them.

It has been found that, to varying degrees, all epoxy adhesives absorb some moisture and are somewhat vapor-permeable, providing another potential means to deliver moisture to the bond interface. This tendency is most pronounced with epoxies that cure at low temperature (*3), i.e., those most likely to be used in homebuilt applications.

In apparent contradiction to all of the above is the experience of sailplane designer Dick Schreder, who had probably experimented more with at-home bonding than anyone else in the homebuilt community.

Schreder writes, "... When we first started using adhesive bonding we quickly discovered that all of the products we tested were too brittle, too weak, non waterproof or promoted corrosion. We finally discovered Hysol EA9410, and have not had any problems in the ensuing 13 years.... I have about 1000 flying hours on the prototype HP-18, which has bonded wings and tail surfaces. All control surfaces and flaps are bonded with no rivets in ribs or trailing edges. 16 foot long flaps on each side of the fuselage are deflected 90 degrees from the inboard ends, and limit terminal velocity to 100 mph in vertical approaches to landing. Although such maneuvers scare observers on the ground, none of the 4 foot flap sections have come unglued to date."

Schreder's wings use 3/8 inch foam ribs placed on 4 inch centers with additional spanwise foam stringers, all of which become bonded area, resulting in light delaminating loads. As Bill Williams, International Sales Manager for all Hysol aerospace products, noted, this construction method could lead to false confidence in bare metal bonds. The foam has low tensile and shear strength, something less than 80 psi. By contrast, Hysol EA9410 has a potential peel and shear strength of 3500 psi and 4500 psi on an aluminum substrate, for a strength ratio of at least 50.1. So, even if the aluminum/adhesive bond strength were degraded by 95%, it would still exceed the load-carrying capacity of the foam, allowing the loss of strength to go unnoticed. Unfortunately, there is no way of knowing if, or when, the critical point will be reached.

Hysol EA9410 is well known and highly respected by most industry people with whom I've spoken, and it may well give the best possible advantage in bonds to surfaces prepared at home. However, it is widely recognized that bond durability is always a function of many factors, of which the adhesive used is only one. (*1, 2, 7) At best, a particularly good adhesive might delay the consequences of inadequate surface preparation, but it can never make up for it.

None of this is intended to imply that **all** aircraft containing bare metal bonds are doomed to failure during their service lives, for such failures are always a matter of statistics. I am concerned that, with the increasing use of aluminum bonding in popular kits, the statistical base is rapidly expanding. Also, as the fleet of homebuilts with inadequately prepared bonds ages, the chance of problems increases for each aircraft. If we continue on the present course, we may well experience a rash of problems that will not only jeopardize our safety, but will devastate the already shaky reputation of aluminum bonding.

INDUSTRY SOLUTIONS

Before describing the solution that finally turned me into an advocate of aluminum bonding for homebuilts, a brief description of the industry's experience is in order.

Early efforts in the aerospace industry to develop a suitable oxide structure resulted in acid etching process involving complex chemical and environment processes involving complex chemical and environmental controls such as FPL Etch, commonly used in its early form from the mid-50s to about 1973. (*4)

The part to be bonded is immersed in a hot (160° F) aqueous solution of sulfuric acid and sodium dichromate. First, the part is etched bright to remove all surface contaminats and oxides. The next step is somewhat controversial, but one study suggests that mixed electrical potential within the molecules of the metal and between the metal and the solution cause a self-anodize to occur, which builds up an exceedingly thin (400 A) and well-attached oxide layer. (*5) Later an equilibrium between new oxide formation and oxide dissolution is reached, after which the part is removed from the solution and rinsed.

When it was found that solution contamination could result in bond failures, a more carefully controlled version of the same process, known as Optimized FPL etch (1973-present) was substituted.

As late as 1975 the Optimized FPL Etch, used in conjunction with various adhesive systems, remained the industry standard surface preparation for primary bonded structures. However, a study by A. W. Bethune of Boeing, published in 1975, contains the following, "... we have 250° F curing modified epoxy adhesives that have exhibited sporadic disbonding not consistently relatable to model, location, or unique service conditions... service experience with bonded structures has been varied, creating a `confidence gap' and `the mixed' performance of the adhesives in service...has created strong reservations in the minds of designers and less-than-enthusiastic acceptance by customers."

I was surprised to learn that an industry pace-setter like Boeing was so recently experiencing this sort of difficulty, but it serves to illustrate the insidious nature of problems associated with aluminum bonding.

Exhaustive analysis of in-service disbonds and the development of new test techniques, which gave much better service correlation than had previously been possible, revealed some classic characteristics of bond failures:

- Disbonds tend to occur in areas of low bondline stress.
- There is always some moisture acess, usually around fasteners.
- Disbonds grow progressively, defined by successive moisture stain "rings"
- Failures are invariably adhesive. They occur between the adhesive/primer and the aluminum, rather than within the adhesive system itself.
- Disbonds occur due to the mechanical fracture of the environmentally weakened (perhaps through hydration) oxide.

Note that nothing in the above would exclude homebuilts, and that these failures were occurring in spite of a sophisticated surface pre-treatment. It has been the universal opinion of the many industry people I've consulted, ranging from bonding facility technicians to veteran researchers with decades of hands-on experience with bonding problems, that difficulties will be vastly more common and severe with bare metal bonds prepared in home shops.

Industry experts concluded that the variables inherent in the Optimized FPL Etch were responsible for the mixed performance experienced in service. This, is turn, led to the conclusion that while suitable for secondary structure, it was not adequate for primary structural applications.

The need for a more durable oxide structure led Boeing to the development of an additional process called Phosphoric Acid Anodize, or PAA (1975-present). For primary structure, the parts are first etched using the FPL Etch, or any of several other deoxidizing etching processes.

Next, parts are immersed in a phosphoric acid solution (at about 80 degrees F), and external voltage (10v) applied to induce the oxide to grow substantially in thickness (to about 4000 A) and roughness.

The PAA process yields consistent results, and greatly enhanced resistance to environmental degradation.

Another process, called Chromic Acid Anodize, is sometimes used in lieu of PAA. This additional process has resulted in literally trouble-free service. As a result, aluminum bonding has regained wide acceptance in the industry.

Regardless of the surface pre-treatment, it has been routine practice to apply a special corrosion-inhibiting adhesive primer (CIAP) before bonding. These primers typically require precise thickness control and careful oven curing. Serving many important functions, this prime coat protects the oxide structure from water, has a tremendous chemical affinity for the epoxy adhesives, and is an excellent general corrosion inhibitor on a continuing basis for exposed portions of bonded parts. Once thus primed, the parts can be handled, stored indefinitely and bonded at any later time after nothing more than a solvent-wipedown. This crucial element is sorely lacking in the scuff/wipe method.

AS FOR THE HOMEBUILTS...

The processes described above are not generally available to the homebuilder. The necessary facilities are complex and costly, involve precise chemical and environmental controls and the handling of large quantities of hazardous substances. Such facilities are possessed only by airframe manufacturers and a few of their sub-contractors.

Composites expert Marshall contends that suppliers of homebuilt kits requiring aluminum bonding should supply the parts already processed and primed, since they would be dealing in sufficient quantities to sub-contract the treatment economically. While this certainly represents the ultimate solution for future kit builders, the 1500 or so of us with projects underway, as well as those building from plans in the future, still have a problem. Because I feel that the "home" should be kept in homebuilts whenever possible, my first line of inquiry naturally focused on an at-home solution.

This raises the question: Is there a means of preparing aluminum at home that will consistently result in reliable bonds? Or, how much room for compromise exists in meeting the five surface-prep criteria stated earlier? This is a tricky question!

The high specifications for the bonded joints of a jet transport pretty much dictate the use of oven-curing high-strength epoxies with wide-service temperature ranges. To match the adhesives' performances, the CIAP (primer) must similarly be of the oven-curing type. In any event, the ultimate objective is to create a "chain," consisting of: Base Metal-Oxide Structure-CIAP-Adhesive, containing no weak links. A properly bonded joint, when destructively tested during or after exposure to a severe environment, will **always** fail cohesively. This means that traces of adhesive will be found completely covering both surfaces.

Clearly, bonds in homebuilts should similarly have no "weak links," and should also fail cohesively when subjected to the above test. However, reasoning that the lesser-strength requirements of homebuilts should allow some room for compromise, an earlier version of this article, completed in January, concluded that interested homebuilders should work together to find a verifiably reliable at-home solution.

My hope at that time was that we could produce a similar "chain" containing no weak links, but with "smaller" links. Clearly we could not expect to produce an oxide structure of our own design at home, but I reasoned that if we "minimized" the oxides using some etch, and quickly followed this with an R. T. Curing primer compatible with the adhesive, we might get sufficiently durable bonds. I even found an etch and a primer that showed promise, as well as a simple test method that has been shown to reliably predict the long-range durability of the bonding method being tested.

This all sounded fine, but the feedback I received on the article indicated a major obstacle that can be described by a single word: - water. It seems that the only effective means of removing oxides and contaminants from the surface of aluminum is by using some powerful acid etch (no fun to handle), which must inevitably be water-rinsed. By all accounts, the bare aluminum exposed by the etch is notoriously sensitive to contamination, and garden-variety water is always loaded with dissolved minerals and contaminants, few of which can easily be filtered out. Thus, all rinse operations at bonding facilities use only highly purified, de-ionized water.

Even if you somehow managed to rinse the part without contaminating it, it must still be thoroughly dried before priming, and of course the surface will be oxidizing during the drying period. You are likely to end up with a contaminated surface that has a nice new layer of poorly attached oxides to bond to! In the PAA process, by contrast, a highly durable oxide structure will have been built up prior to the need for drying.

Although early indications had pointed toward prohibitive cost, I decided to send copies of my article and a parts list for my wings to several aerospace subcontractors. Interestingly, the greatest difficulty was convincing them of the magnitude of the homebuilt movement, for the industry seems largely oblivious to its size and rate of expansion.

Eventually I received a call from George Dalley, Marketing Manager for Panel-Air in Costa Mesa, California, who told me that his company was interested in serving the homebuilt market. The estimate he gave for processing my parts, which consist of two skins (5 feet x 12 feet), two spars (14 feet each) and about 70 stamped ribs and miscellaneous other parts, was $500.

On the appointed day I made the 500 mile trek, and was greeted by five technicians eager to work on my parts. The parts were wiped down with solvent and carefully clamped to an aluminum rack that provides support and acts as a ground connection during the anodize step. From this point, surgical cleanliness was the rule.

The impressive PAA installation consisted of seven tanks measuring about 3 feet x 20 feet x 14 feet deep, and I watched as the parts were successively lowered into the tanks for alkaline de-greasing, rinsing, FPI, etch, rinsing, phosphoric acid anodize, rinsing and drying.

Other than looking clean, there was little visible evidence of change, although at flat angles the anodized surface refracted light in colorful hues. (This is used as a means of testing the anodize.) Next, a thin coat (0.004 mm) of CIA primer was applied, followed by oven curing for one hour at 250 degrees F. After cooling, the parts were wrapped for protection and returned to me with a statement certifying that they had been processed per Boeing specification BAC 5555.

A tour of Panel-Air, which also has a complete machine shop, a parts and materials warehouse, and is an FAA Certified Repair Station, showed that it is well equipped to serve kit vendors.

Several special parts like extrusions or stamped parts could be kept on hand, and processed for the kit vendor as needed. Common materials like skins could be taken directly from Panel-Air's own stock.

REFERENCES

*1. Corey McMillan, *Surface Preparation—the Key to Bondment Durability*, Advisory Group for Aerospace Research & Development, 7 Rue Ancelle 92200 Neuilly Sur, Seine, France, 1978.

*2. Murray H. Kuperman, Manager, Engineering Standards & Services, United Airlines. Correspondence to Otis Holt, 9/13/81.

*3. A.W. Bethune, *Durability of Bonded Aluminum Structure, SAMPE Journal*, Vol. 11, No. 3, Jul./Aug./Sept. 1975.

*4. H.W. Blackner and N.E. Schowalter, *A Study of Methods for Preparing Clad 2S-T3 Aluminum Alloy Sheet Surfaces for Adhesive Bonding*, Forest Products Laboratory Report No. 1813, May 1950.

*5. A.W. Smith, Surface Oxide on Etched Aluminum, J. Electro-chemical Society, Solid-State Science and Technology, Nov. 1973.

*6. Andrew C. Marshall, from correspondence to Otis Holt, 8/21/82 and 9/18/82.

APPENDIX 5

Table 5-1. Properties of a Few Carbon Fiber Types

Fiber Manu-facturer	Fiber Designation	No. of Filaments (thousands)	Tensile Strength (ksi)	Tensile Modulus (msi)	Fiber Diameter (microns)	% of Elonga-tion	Fiber Weight (gm/cc)
PAN BASED FIBER							
Toray	T300	1,3,6,12	512	33	7	1.5	1.8
Amoco	T300	1,3,6,12	530	33	7	1.4	1.8
Hercules	AS4	3,6,12	560	33	7	1.6	1.7
Toray	T300J	1,3,6,12	611	33	7	1.8	1.8
Toray	T700S	12	711	33	7	2.1	1.8
Amoco	T650/35	3,6,12	660	35	7	1.8	1.8
Hercules	AS4D	12	600	35	7	1.6	1.8
Hercules	IM2A	12	700	40	5	1.6	1.8
Hercules	IM4C	12	600	40	5	1.4	1.7
Hercules	IM7	6,12	770	40	5	1.7	1.8
Hercules	IM8	12	700	40	5	1.6	1.8
Amoco	T40	12	820	42	5	1.8	1.8
Amoco	T650/42	12	700	42	5	1.7	1.8
Toray	M30	1,3,6,12	570	43	5	1.3	1.7
Toray	M30SC	18	800	43	5	1.3	1.7
Toray	T800	6,12	800	43	5	1.9	1.7
Toray	T1000G	12	943	43	5	2.1	1.8
Hercules	HMS6C	12	600	52	5	1.0	1.8
Amoco	T50	3,6,12	420	57	5	0.7	1.8
Amoco	T55X	6,12	630	55	6	1.2	1.8
Toray	M40J	6,12	640	55	5	1.2	1.8
Toray	M50J	6	570	69	5	0.8	1.9
Toray	M60J	3,6	570	85	5	0.7	1.9
PITCH BASED FIBER							
Amoco	P-25	2,4	200	23	10	0.9	1.9
Amoco	P-55	2,4	300	66	10	0.5	2.0
Amoco	P-75	2	300	75	10	0.4	2.0
NGF	XN-50A	0.5,1,2,4	560	75	10	0.7	2.15
NGF	YSH-50A	1.5,3,6	560	75	7	0.7	2.10
NGF	XN-70A	0.5,1,2	530	105	10	0.5	2.15
NGF	YSH-70A	1.5,3,6	530	105	5	0.7	2.15
NGF	YS-90A	3,6	510	130	7	0.7	2.15
Amoco	P-100	2	350	110	10	0.3	2.16
Amoco	P-120	2	350	120	10	0.3	2.17
NOC	XN-85A	2	530	120	10	0.4	2.15
Amoco	K-800X *	2	400	125	10	0.3	2.15
Amoco	K-1100 *	2	450	140	10	0.3	2.2

* Used for high thermal conductivity structures

NOTE: All of the above properties are as represented by the manufacturer's data sheets.

Table 5-2. Carbon Fabrics Manufactured by Textile Products, Inc.

STYLE	YARN	COUNT yarns/in.	WEAVE style	WEIGHT oz./sq.yd.	THICKNESS in.
4110	1M7-6K	11.5 X 11.5	CFS	5.84	.014
4114	T300--6K	12 X 12	5HS	11.05	.027
4116	AS4-6K	10.5 X 10.5	5HS	11.05	.027
4133	G30-500-3K	12.5 X 12.5	PLAIN	5.84	.013
4134	T300-1K	24 X 24	5HS	3.77	.011
4147	AS4-3K	11.5 X 11.5	PLAIN	5.85	.013
4160	T650-35-3K	12.5 X 12.5	PLAIN	5.84	.012
4165	T300-3K	12.5 X 12.5	TWILL	5.93	.014
4176	1M7-6K	16 X 16	5HS	8.38	.020
4245	G30-500-3K	24 X 23	8HS	11.08	.024
4562	T300-12K	10.5 X 10.5	TWILL	19.31	.038

NOTE: 1.0 oz./sq. yd. = 33.3937 gm. sq. meter

Table 5-3. Comparison of Typical High Modulus Aerospace Grade Graphite Fiber

Fabric Types	Fiber	Fiber Manuf.	Fiber Tens. Modulus (msi)	Tow Size (K)	Fiber Conductivity (W/m°K)	FAW (g/m2)	Fabric Weight (oz/yd)	Fabric Width (m)	Fiber Diam. (microns)
Plain or Satin	M40J	Toray	54	6 K	38	>200	>5.6	-	5
Plain or Satin	M46J	Toray	63	6 K	42	>200	>5.6	-	5
SF-50A-75	XN-50A	NGF	75	0.5 K	180	75	2.2	0.5	10
SF-50A-105	XN-50A	NGF	75	1 K	180	105	3.1	0.5	10
5HS-50A-140	XN-50A	NGF	75	1 K	180	105	4.2	1.4	10
PF(S)-50A-140	XN-50A	NGF	75	1 K	180	140	4.2	1.0	10
PF(S)-50A-150	XN-50A	NGF	75	2 K	180	150	4.5	1.0	10
PF(S)-YSH50A-200	YSH-50A	NGF	75	3 K	120	200	5.6	1.0	7
PF(S)-YSH60A-110	YSH-60A	NGF	92	3 K	180	110	3.3	1.0	7
Open PF	M55J	Toray	78	6 K	65	175	5.2	1.0	5
Satin	M60J	Toray	85	3 K	75	150	4.5	1.0	5
Satin	P100	Amoco	105	2 K	500	175	5.2	1.0	10
SF-70A-75	XN-70A	NGF	105	0.5 K	320	75	2.2	0.5	10
SF-70A-105	XN-70A	NGF	105	1 K	320	105	3.1	0.5	10
5HS-70A-140	XN-70A	NGF	105	0.5 K	320	140	4.2	1.0	10
PF(S)-70A-150	XN-70A	NGF	105	3 K	250	300	9.0	1.0	10
5HS-YSH70A-300	YSH-70A	NGF	105	3 K	250	300	3.7	1.0	10

NOTES:

1 "SF" indicates a plain weave which is "Spread", making each yarn thinner and wider.
2. "5HS" indicates a 5-Harness Satin Weave.
3. "PF" indicates a Plain Weave Fabric.
4. "Open PF" indicates a plain weave with substantial openings between adjacent yarns.
5. This table was prepared by Thomas G. Wong, a consultant on materials for space vehicles, for use by a designer considering the available choices of materials for the facings of Solar Array Panels and Antenna Dish Structures.

APPENDIX 6

Table 6-1. Sandwich Panels and Cargo Liners Available from M.C. Gill Corporation

Some of the sandwich panels and cargo liners listed below are maintained in inventory, and can be shipped immediately from stock. All panels are used in flooring or cargo liners in various locations in many commercial aircraft.

M.C. GILL Product Designation	DIMENSIONS	CONSTRUCTION	COMMON USES
Gillfloor™ 4417 Panel	0.400" x 48" x 144" Facings 0.030"/0.030"	Unidirectional fiberglass reinforced epoxy faces bonded to 12 pcf Nomex® honeycomb core with modified epoxy adhesive.	Aircraft flooring suitable for passenger compartment underseat, aisles, entries, galleys and cargo compartments depending on Type. Qualified to Boeing BMS 4-17, Type IX.
Gillfloor™ 4417A Panel	0.400" x 48" x 144" Facings 0.020"/0.020"	Unidirectional fiberglass reinforced epoxy faces bonded to 10 pcf Nomex® honeycomb core with modified epoxy adhesive.	Aircraft flooring suitable for passenger compartment underseat, aisles, entries and galleys. Qualified to Boeing BMS 4-17, Type VI.
Gillfab™ 4505 Panel	0.374" x 48" x 144" Facings 0.020"/0.020"	Unidirectional carbon/woven glass reinforced phenolic facings bonded to 9 pcf Nomex®. honeycomb core floor panel.	Passenger and cockpit flooring for Airbus single aisle and long range aircraft. Qualified to Airbus technical specification 5360 M1M 000600, Type PC3.
Gillfloor™ 4509 Panel	0.390" x 48" x 144" Facings 0.015"/0.015"	Unidirectional carbon reinforced phenolic facings bonded to 8 pcf Nomex® honeycomb core.	Passenger compartment flooring. Qualified to McDonnell Douglas drawing 7954400, all types.
Gillfab™ 4522 Panel	0.374" x 48" x 144" Facings 0.020"/0.015"	Woven fiberglass cloth reinforced phenolic facings bonded to 8 pcf Nomex® honeycomb core.	Containerized cargo compartment flooring for Airbus. Qualified to Airbus technical specification 5360 M1M 000500, (CCC1).
Gillfloor™ 5007B Panel	0.400" x 48" x 144" Facings 0.040"/0.020"	Woven fiberglass cloth reinforced polyester facings bonded to 9 pcf end-grain balsawood core.	Aircraft replacement flooring in high traffic areas. Qualified to United Airlines SHE 2902.
Gillfloor™ 5007C Panel	0.400"x48"x144" Facings 0.045"/0.025" or 0.045"/0.045"	Woven fiberglass cloth reinforced polyester facings with wear-resistant overlay on one or both sides, bonded to 9 pcf end-grain balsawood core.	Bulk and containerized cargo compartment replacement flooring. Proprietary to the M.C. Gill Corp.
Gillfab™ 5040 Panel	0.400"x 48" x 144" Facings 0.020"/0.012"	2024-T3 aluminum faces bonded to 9 pcf end-grain balsawood core with heat setting elastomeric adhesive.	Passenger and cargo compartment replacement flooring. Used as general purpose panel to replace plywood.
Gillfab™ 5042B Panel	0.390" x 48" x 144" Facings 0.015"/0.012"	7075-T6 aluminum faces bonded to 9 pcf end-grain balsawood core with modified epoxy adhesive.	Normally used in commercial aircraft flooring, galley panels, cargo containers and bulkheads. Qualified to McDonnell Douglas specifications S3932193, S3932195 and S4931863.
Gillfab™ 5242 Panel	0.390" x 48" x 144" Facings 0.040"/.012"	Top facing 0.020" epoxy/fiberglass bonded to 0.020" 2024-T3 aluminum; bottom facing 0.012"; facings bonded to 9 pcf end-grain balsawood core with epoxy adhesive.	Cargo flooring in DC-9, MD-80 aircraft. Qualified to McDonnell Douglas Drawing S00096, Rev. C.
Gillfab™ 5424 Panel	0.400" x 48" x 144" Facings 0.015"/0.015"	Unidirectional fiberglass reinforced epoxy faces bonded to 8.5 pcf aluminum honeycomb core with modified epoxy adhesive.	Passenger compartment flooring in 737 and 757 aircraft. Qualified to Boeing 4-23, Type II.

APPENDIX 7

Murphy's Law

Ever since mechanics and engineers have been working with mechanical devices, they have often been perplexed by the whims and jokes that the real world plays on them.

At first, it applied only to mechanical situations but it now seems as if all the arts and sciences have developed their very own versions of Murphy's Law. The version that seems most appropriate, and one which was posted on the wall of the laboratory at the Hexcel Corporation for many years, is approximately those shown below.

MURPHY'S LAW

1. Anything that can go wrong will go wrong.
2. If the simultaneous occurrence of two entirely unrelated events would somehow result in a very serious problem, that is exactly what will occur.
3. It is against the Laws of Nature to have things go the way they are supposed to go.
4. Mother Nature is a bitch.
5. Occasionally Murphy's Laws themselves are subject to Murphy's Law, and to everybody's astonishment something very productive occurs.